Cement Manufacturer's Handbook

by

Kurt E. Peray

Chemical Publishing Co., Inc.

New York, N.Y.
1979

ISBN 0-8206-0245-0

Chemical Publishing Co., Inc.

Printed in the United States of America

PREFACE

With this book, the author has intended to fill the need for a handy reference guide for cement plant engineers, supervisors, and managers.

This work contains the necessary engineering formulas which represent the basic tools for gaining a better understanding of cement manufacturing technology. Mathematical formulas have been purposely kept simple with a minimum of written text to conserve an engineers time and to make this book available for use to the widest possible readership.

Little or no mention is made about the result one can expect from any of the multitude of formulas presented. To do so would not serve a useful purpose and would defeat the objectives set forth for this book. None of the results obtained are universally applicable as is true for all manufacturing facilities. The process of manufacturing cement is one of the most difficult and dynamic processes known. Therefore, it would be a fallacy to assume that because a certain equipment, design, or method has proven itself in some plants that it would be equally successful in another. Nowhere is this situation more pronounced than in the rotary kiln operation.

There is no substitute for experience. But, experience combined with a sound mathematical understanding and knowledge of the process is a goal every cement plant engineer should strive for. If this book generates interest and gives the reader more satisfaction on his or her job, its writing will have been justified. But it is hoped, that it will do more than that.

* * *

Special appreciation is extended to my family for their sacrifices without which this book would not have become a reality. For their understanding and patience I'm dedicating this book to

SONJA, DANIELA, AND MARCEL

* * *

INTRODUCTION

The process of manufacturing cement requires a knowledge of many sciences such as chemistry, physics, thermodynamics, and physical chemistry. It takes years to acquire a fundamental knowledge of the cement manufacturing process and this learning process tends to be never ending. There is always something new in this industry: new types of equipment, new operating techniques, and conditions that require great effort to bring about change. Failure to accept this as a requirement of the job could lead to stagnancy instead of progress.

Unfortunately, for anyone that has chosen a career in the cement industry, this knowledge cannot be acquired in a classroom or behind a desk. Instead, to become familiar with the process, one must acquire experience out in the field. For a manager or engineer this means he periodically has to don his overalls and spend time on the firing floor, the grinding department, the packhouse, and the quality control laboratory. By doing this, he must not shy away from the possibility that he might return to the office in a state more fitting for a chimney cleaner than a member of the plants management staff.

In the course of 23 years, the author has been working in almost all departments of a cement plant and has been fortunate to gather experience as an hourly employee as well as a member of the plants management staff. During these times, a sore back from loading cement sacks on trucks, clinker dust in nostrils and ears from operating a kiln, inflamed eyes from cement and kiln dust, a few minor burns from hot clinker, a lot of sleepless nights, call-outs, and 16-hour work days were common occurrences.

It is all part of the learning process.

It is hoped that these remarks do not give the newcomer to the industry the false impression that all cement plants are terribly dirty places to work in. They really aren't, they are just slightly different from other industrial processes and some time is required to get used to them. There is however a much more positive aspect to embarking on a career in the cement industry. Making a living in the cement industry, as a manager, engineer, supervisor, or hourly worker, is a job that seldom becomes monotonous or boring. This is an interesting technical field to work in and is always full of the unexpected. It takes a special kind of individual who can tackle new problems head-on. It is a credit to the cement industry that it has so many individuals that can salvage an apparently hopeless situation and keep, figuratively speaking, the train on the track. The author himself has observed specialists in their own right making a production facility continue to produce cement when others had given up years ago saying that the particular equipment was long overdue for the scrap pile. And at other times, workers and supervisors, almost beyond their call of duty, have proven they can repair a piece of equipment and get it back on line within a time frame other industries would consider impossible. These are the unsung heroes in the cement industry, those who are just doing their job and whose names usually never appear in the trade literature. In part, the author dedicates this book to these individuals.

There really is no way for a book to teach the uninitiated this kind of a work experience. There are just too many unknowns and variables that enter into the decision making process about how to handle a given situation. It is therefore a fallacy to assume that this book provides an individual with everything he has to know about the cement manufacturing process. The aim of this book, however, is to provide the foundation upon which an individual can build his experience and technical know-how. The author has attempted to compile the technical information that is considered necessary to give the reader a good background of the process.

It is not uncommon to observe an engineer spending four hours in preparation of a test, one hour for the actual test in the field, two hours for calculations, and two days in compiling the results and writing the report. In many chapters of this book, work sheets are provided which an engineer can copy thus saving him valuable time in this overall endeavor.

Since the majority of the formulas in this book are presented both in the English and metric systems, the engineer has a tool available that makes the transition to the new system easier for him. The only caution the author must give is that the reader should make a habit of ascertaining the appropriate formula in the correct system of units for his work or project. Included at the end of this book are extensive conversion tables that allow the reader to become familiar with all three systems: the English, the metric, and the International System of Units.

The author would like to see a college or university that would establish a school of "Cement Manufacturing Technology" here in the United States. Such an institution would enable our industry to develop the required pool of new engineers needed to maintain a progressive technological growth in the North American industry. It is the authors opinion that such a school could contribute a great deal toward making the U.S. cement industry less dependent upon foreign technology. There are many unique processes that were invented and developed by the U.S. cement industry. Perhaps in the future we can again take a leading role in improving and advancing the technology of making cement. But to do so requires a financial commitment and a great deal of effort from all of us.

Kurt E. Peray
Dallas, Texas

CONTENTS

Part I Cement Chemistry

Part II Burning

Part IV Engineering Formulas

Part VI Appendix

PART I

CEMENT CHEMISTRY

Chapter 1

QUALITY CONTROL FORMULAS

1.01 Ignition Loss

Ignition loss is usually determined by tests in a laboratory furnace. It can also be calculated from the chemical analysis of the kiln feed by the following formula:

$$\text{Ignition loss} = 0.44\ CaCO_3 + 0.524\ MgCO_3 + \ldots\ldots + \text{combined } H_2O + \text{organic matter.}$$

1.02 Silica Ratio

$$SR = \frac{SiO_2}{Al_2O_3 + Fe_2O_3}$$

Large variations of the silica ratio in the clinker can be an indication of poor uniformity in the kiln feed or the fired coal. Changes in coating formation in the burning zone, burnability of the clinker, and ring formations within the kiln can often be traced to changes of the silica ratio in the clinker. As a rule, clinker with a high silica ratio is more difficult to burn and exhibits poor coating properties. Low silica ratios often lead to ring formations and low early strength (3–7 days) in the cement.

1.03 Alumina-Iron Ratio

$$A/F = \frac{Al_2O_3}{Fe_2O_3}$$

Clinker with a high alumina-iron ratio, as a rule, produce cement with high early strength (1 to 3 days) but makes the reaction between the silica and calcium oxide in the burning zone more difficult.

1.04 Lime Saturation Factor

This factor has been used for kiln feed control for many years in Europe and only recently has also found acceptance by American cement manufacturers. When the lime saturation factor approaches unity, the clinker is difficult to burn and often shows excessive high free lime contents. A clinker, showing a lime saturation factor of 0.97 or higher approaches the threshold of being "overlimed" wherein the free lime content could remain at high levels regardless of how much more fuel the kiln operator is feeding to the kiln.

If $A/F = > 0.64$

$$LSF = \frac{CaO}{2.8\ SiO_2 + 1.65\ Al_2O_3 + 0.35\ Fe_2O_3}$$

If $A/F = < 0.64$

$$LSF = \frac{CaO}{2.8\ SiO_2 + 1.1\ Al_2O_3 + 0.7\ Fe_2O_3}$$

1.05 Hydraulic Ratio

This index is very seldom used any more in modern cement technology for kiln feed control.

$$HR = \frac{CaO}{SiO_2 + Al_2O_3 + Fe_2O_3}$$

1.06 Percent Liquid

Clinker, when burned at a temperature of 2642°F., has the following liquid content:

$$\text{Percent liquid} = 1.13\ C_3A + 1.35\ C_4AF + MgO + \text{Alkalies}$$

1.07 Burnability Index

This is an indicator of the ease of burning for a given clinker. The higher the index number, the harder the clinker is to burn.

$$BI = \frac{C_3S}{C_4AF + C_3A}$$

1.08 Burnability Factor

The burnability factor is used as a guideline for the kiln operator to show if a given clinker is easier or harder to burn. Higher burnability factors yield a clinker that is harder to burn. Conversely, lower factors make the clinker easier to burn.

$$BF = LSF + 10\ SR - 3\ (MgO + \text{Alkalies})$$

(find *LSF* in **1.04** and *SR* in **1.02**.)

1.09 Bogue Formulas for Clinker and Cement Constituents

For a cement chemist, these formulas are the most important and frequently used indicators of the chemical properties of a cement or clinker. The constituents calculated by these formulas, however, are only the potential compositions when the clinker has been burned and cooled at given conditions. Changes in cooling rate or burning temperature can modify the true constituent composition to a considerable extent.

a) Bogue Formulas for Cement Constituents

If $A/F = >0.64$

$$C_3S = 4.071CaO - (7.602SiO_2 + 6.718Al_2O_3 + 1.43Fe_2O_3 + 2.852SO_3)$$

$$C_2S = 2.867SiO_2 - 0.7544C_3S$$

$$C_3A = 2.65Al_2O_3 - 1.692Fe_2O_3$$

$$C_4AF = 3.043Fe_2O_3$$

If $A/F = <0.64$

$$C_3S = 4.071CaO - (7.602SiO_2 + 4.479Al_2O_3 + 2.859Fe_2O_3 + 2.852SO_3)$$

$$C_2S = 2.867SiO_2 - 0.7544C_3S$$

$$C_3A = 0$$

$$ss(C_4AF + C_2F) = 2.1Al_2O_3 + 1.702Fe_2O_3$$

b) Bogue Formulas for Clinker Constituents

When appreciable amounts of SO_3 and Mn_2O_3 are present in the clinker, the values of the chemical analysis have to be recalculated to take into account the amount of CaO that has been combined with SO_3, the amount of free lime present and the Mn_2O_3.

The values to be used in the Bogue formulas are:

$$Fe_2O_3 = Fe_2O_3 + Mn_2O_3$$

$$CaO = CaO - \text{free } CaO - (CaO \text{ combined with } SO_3)$$

To find the amount of CaO that is combined with SO_3 as $CaSO_4$ proceed as follows:

Step 1 If $(K_2O/SO_3) = < 1.176$ then not all of the SO_3 is combined with K_2O as K_2SO_4

$$SO_3 \text{ in } K_2O = 0.85\ K_2O$$

Step 2 Calculate SO_3 residue

$$SO_3 - SO_3\ (\text{in } K_2O) = SO_3\ (\text{remaining})$$

If $[Na_2O/SO_3\ (\text{remain.})] = < 0.774$ then not all of the remaining SO_3 is combined with Na_2O as Na_2SO_4.

$$SO_3 \text{ in } Na_2O = 1.292\ Na_2O$$

Step 3 Calculate the amount of CaO that has combined with the SO_3 as $CaSO_4$

$$CaO\ (\text{in } SO_3) = 0.7\ [SO_3 - SO_3\ (\text{in } K_2O) - SO_3\ (\text{in } Na_2O)]$$

Having determined the appropriate values for the CaO and Fe_2O_3, one can then proceed to calculating the potential clinker constituents by using the previously given Bogue formulas. When the Bogue formulas are used for kiln feed compositions, keep in mind that the coal ash addition, dust losses, and alkali cycles can alter the final composition of the clinker. Also use the analysis on a "loss free" basis in the calculations of the constituents.

1.10 Total Carbonates

Total carbonates are usually determined analytically by the acid-alkali titration method. They can also be calculated from the raw (unignited) analysis as follows:

$$TC = 1.784\ CaO + 2.09\ MgO$$

1.11 Total Alkalies as Na_2O

The total alkali content in terms of sodium oxide is calculated from the loss free analysis:

$$\text{Total as } Na_2O = Na_2O + 0.658\ K_2O$$

1.12 Conversion of Raw Analysis to Loss Free Basis

$$O_i = \frac{O_r}{100 - L} 100$$

where

O_r = percent of oxide (by weight) on a raw basis

O_i = percent of oxide (by weight) on loss free basis

L = percent loss on ignition (by weight)

1.13 Conversion of Kiln Dust Weight to Kiln Feed Weight

Dust collected in a precipitator or baghouse of a kiln shows a different loss on ignition than the kiln feed because it has been partially calcined. For inventory control purposes and in some kiln operating studies it is often necessary to express the weight of dust in terms of equivalent feed weight.

$$w_e = \frac{(w_d)(1 - L_d)}{1 - L_f}$$

where

w_e = weight of dust in terms of feed

w_d = actual weight of dust

L_d = percent ignition loss, dust (decimal)

L_f = percent ignition loss, kiln feed (decimal)

1.14 Calculation of Total Carbonates from Acid-Alkali Titration

This method is only applicable when the MgO content of the sample is known. Values from the raw (unignited) basis are used for the calculation.

$CaCO_3 = 1.66791\,(a - 1.48863\ MgO)$

$MgCO_3 = 2.098\ MgO$

$TC = MgCO_3 + CaCO_3$

$CaO = 0.93453\,(a - 1.48863\,MgO)$

a = apparent total lime content from titration

1.15 Percent Calcination

Kiln feed or dust samples taken at any location of the kiln are often investigated for the apparent degree of calcination the sample has undergone.

$$C = \frac{(f_i - d_i)}{f_i} 100$$

where

C = apparent percent calcination of the sample

f_i = ignition loss of the original feed

d_i = ignition loss of the sample

PROBLEMS AND SOLUTIONS

Problems and examples shown in this chapter and all subsequent chapters are arranged in the same sequence as the formulas are presented in the chapter.

1.01 A kilnfeed mix contains 78.5 percent $CaCO_3$, 1.2 percent $MgCO_3$ and an estimated 0.4 percent combined H_2O and organic matter. What is the ignition loss on this mix?

Ignition loss = $(0.44 \times 78.5) + (0.524 \times 1.2) + 0.4 = 35.6$ percent *(ans.)*

1.02 It is desired that a given kiln feed shows a silica ratio of 2.75. What must the Al_2O_3 content be if the Fe_2O_3 remains constant at 2.95 and the SiO_2 at 22.45?

$$Al_2O_3 = \frac{22.45 - (2.95 \times 2.75)}{2.75} = 5.21 \text{ percent} \quad (ans.)$$

1.03 The Al/Fe ratio is desired to be 1.80 and the Fe_2O_3 is to be 2.75 in a given mix. What must the Al_2O_3 content be?

$Al_2O_3 = 1.8 \times 2.75 = 4.95$ percent *(ans.)*

1.04 Given a clinker of the following composition:

CaO	SiO_2	Al_2O_3	Fe_2O_3
66.75	22.15	5.96	2.93

What is the lime saturation factor?

First

$$\frac{Al}{Fe} = \frac{5.96}{2.93} = 2.03 \qquad \text{i.e.,} > 0.64$$

Hence,

$$LSF = \frac{66.75}{[(2.8)(22.15)] + [(1.65)(5.96)] + [(0.35)(2.93)]} = 0.916$$

(*ans.*)

1.05 What is the hydraulic ratio for the clinker example given in **1.04**?

$$HR = \frac{66.75}{22.15 + 5.96 + 2.93} = 2.15 \quad (\textit{ans.})$$

1.06 Given a clinker of the following characteristics:

C_4AF	MgO	alkalies
8.97	2.15	0.65

What must the C_3A content be to obtain 25.5 percent liquid in the clinker?

$$C_3A = \frac{25.5 - 12.11 - 2.15 - 0.65}{1.13} = 9.37 \quad (\textit{ans.})$$

1.07 What is the burnability index for a clinker that shows 61.5 percent C_3S, 8.8 percent C_4AF, and 9.8 percent C_3A?

$$BI = \frac{61.5}{8.8 + 9.8} = 3.31 \quad (\textit{ans.})$$

1.08 What is the burnability factor for a clinker of the following characteristics:

$LSF = 92.0, \quad S/R = 2.75, \quad MgO + alk. = 3.15$

$BF = 92.0 + [(10)(2.75)] - [(3)(3.15)] = 110.1 \quad (\textit{ans.})$

1.09 Given a clinker of the following composition:

Si	Al	Fe	CaO	MgO	Mn_2O_3	SO_3	K_2O
21.84	5.15	2.65	66.85	0.85	0.12	1.2	0.6

Na_2O	free CaO
0.15	0.5

What values for the basic oxides have to be used when calculating the compound composition by the Bogue formula?

a) For Fe_2O_3

$$Fe_2O_3 = 2.65 + 0.12 = 2.77 \quad (ans.)$$

b) For CaO

$$\frac{K_2O}{SO_3} = 0.5, \qquad SO_3 \text{ in } K_2O = (0.85)(0.60) = 0.51$$

$$SO_3 \text{ remaining} = 1.2 - 0.51 = 0.69$$

$$\frac{Na_2O}{SO_{3\,rem.}} = \frac{0.15}{0.69} = 0.217$$

$$SO_3 \text{ in } Na_2O = (1.292)(0.15) = 0.194$$

$$CaO \text{ in } SO_3 = 0.7\,(1.2 - 0.51 - 0.194) = 0.347$$

$$CaO = 66.85 - 0.5 - 0.347 = 66.00 \quad (ans.)$$

Hence, the oxide contents to be used are

CaO	SiO_2	Al_2O_3	Fe_2O_3
66.00	21.84	5.15	2.77

1.10 It is desired that a mix contain 79.8 percent total carbonates. What must the CaO content be if the MgO is a constant 1.3 percent on a raw basis?

$$CaO = \frac{79.8 - 2.72}{1.784} = 43.21 \text{ percent} \quad (\textit{ans.})$$

1.11 What is the maximum permissible K_2O content in the clinker if the Na_2O content is 0.18 percent and the total number of alkalies is not allowed to exceed 0.63 percent?

$$K_2O = \frac{0.63 - 0.18}{0.658} = 0.68 \quad (\textit{ans.})$$

1.12 In example **1.10**, what is the CaO content, loss free, when the loss on ignition is 35.7 percent on this mix?

$$CaO_{(\text{loss free})} = \frac{43.21}{100 - 35.7}\, 100 = 67.20 \text{ percent} \quad (\textit{ans.})$$

1.13 A kiln is wasting 5850 lb of dust per hour. The loss on ignition of the dust is 19.5 percent and 35.8 percent on the kiln feed. What weight of kiln feed is wasted on this kiln per hour?

$$w_c = \frac{5850(1 - 0.195)}{1 - 0.358} = 7335.3 \text{ lb/h} \quad (\textit{ans.})$$

1.14 What is the total carbonate content on a sample that shows CaO = 47.0 percent and MgO = 0.84 percent on a raw basis?

$$TC = 1.66791\, [47.0 - (1.48863)(0.84) + (2.098)(0.84)]$$

$$= 79.25 \text{ percent} \quad (\textit{ans.})$$

1.15 To what percent is the kiln dust in example **1.13** above calcined?

$$\text{Percent calcination} = \frac{35.8 - 19.5}{35.8}\, 100 = 45.5 \text{ percent} \quad (\textit{ans.})$$

Chapter 2

KILN FEED MIX CALCULATIONS

2.01 $CaCO_3$ Required to Obtain a Given C_3S in the Clinker

This formula should only be used as a quick reference in times when no other analytical methods, other than the titration method, is available.

$$CaCO_3 \text{ required} = CaCO_{3\,(\text{feed})} + 0.0806\,(A - a)$$

where

A = desired C_3S in clinker
a = existing C_3S in clinker

2.02 Two-Component Mix Calculations

a) To obtain a constant total carbonate content

This method can only be used when the $MgCO_3$ content in the two components is constant.

$$TC = \frac{x\,C_1 + (100 - x)\,C_2}{100}$$

where

x = material A needed (percent by weight)

$100 - x$ = material B needed (percent by weight)
C_1 = TC in material A (percent by weight)
C_2 = TC in material B (percent by weight)
TC = desired total carbonates

b) Percent of each component needed for a desired $CaCO_3$

Use this formula only when the $MgCO_3$ content in the two components is constant.

$$w = \frac{C_f - C_2}{C_1 - C_f} 100$$

where

w = weight of material A needed for each 100 unit weights of material B
C_f = $CaCO_3$ desired in mix
C_1 = $CaCO_3$ in material A
C_2 = $CaCO_3$ in material B

c) To obtain a constant C_3S/C_2S ratio

Insert the values found from the raw material analysis (on the raw basis)

	limestone	*clay*
SiO_2	S_1	S_2
Al_2O_3	A_1	A_2
Fe_2O_3	F_1	F_2
CaO	C_1	C_2
MgO	M_1	M_2
Loss	L_1	L_2

For limestone

$$x = C_1 + 1.4M_1 - (2.3S_1 + 1.7A_1 + F_1)$$

For clay

$$y = 2.3S_2 + 1.7A_2 + F_2 - (C_2 + 1.4M_2)$$

The number of parts (P) limestone required for 100 parts of clay will be:

$$P = \frac{100\,y}{x}$$

The percent (by weight) of the mix is therefore

$$\text{Percent clay} = \frac{100}{100 + P}\,100$$

$$\text{Percent limestone} = \frac{P}{100 + P}\,100$$

d) Formulas for mix corrections

Limestone added to a cement rock to correct mix.

$$\chi = A + \frac{(M - F)(100 + A)}{L - M}$$

where

M = percent $CaCO_3$ desired in mix
F = percent $CaCO_3$ found in mix (before correction)
A = percent limestone already added
L = percent $CaCO_3$ in limestone
χ = corrected percent limestone needed to obtain M.

Clay added to a limestone to correct mix.

$$\chi = B - \frac{(M - F)(100 + B)}{F - C}$$

where

M = percent CaO or $CaCO_3$ desired in the mix
F = percent CaO or $CaCO_3$ found in mix
B = percent clay already added
C = percent CaO or $CaCO_3$ in clay
χ = percent clay needed to obtain M

2.03 Three Component Mix Calculation

a) To obtain a desired LSF and SR

Analysis on Raw Basis

	Material I	*Material II*	*Material III*
SiO_2	S_1	S_2	S_3 .....
Al_2O_3	A_1	A_2	A_3 .....
Fe_2O_3	F_1	F_2	F_3 .....
CaO	C_1	C_2	C_3 .....

Desired *LSF*: KS_k = ...
Desired *S/R*: SM = ...

$a_1 = [(2.8KS_k)(S_1) + 1.65A_1 + 0.35F_1)] - C_1 =$ .....
$b_1 = (2.8KS_k)(S_2) + 1.65A_2 + 0.35F_2) - C_2 =$ .....
$c_1 = C_3 - (2.8KS_k + 1.65A_3 + 0.35F_3) =$ .....
$a_2 = SM(A_1 + F_1) - S_1 =$ .....
$b_2 = SM(A_2 + F_2) - S_2 =$ .....
$c_2 = S_3 - SM(A_3 + F_3) =$ .....

$$\chi = \frac{c_1 b_2 - c_2 b_1}{a_1 b_2 - a_2 b_1} = \quad \ldots\ldots \quad \text{Parts of Material I}$$

$$y = \frac{a_1 c_2 - a_2 c_1}{a_1 b_2 - a_2 b_1} = \quad \ldots\ldots \quad \text{Parts of Material II}$$

1.0 ..... Parts of Material III

u = Sum Total of all Parts	=	
z = 100/u	=	
Percent of Material I χz	=	percent
Percent of Material II yz	=	percent
Percent of Material III 1.000z	=	percent
Total	100	percent

b) To obtain a desired SR and HR

Let the following represent the analysis of the three raw materials (on a raw basis)

	Comp. 1	*Comp.* 2	*Comp.* 3
SiO_2	S_1	S_2	S_3
$Al_2O_3 + Fe_2O_3$	O_1	O_2	O_3
$CaCO_3 + MgCO_3$	L_1	L_2	L_3

Let

$$r = \text{silica ratio desired} = \frac{SiO_2}{Al_2O_3 + Fe_2O_3}$$

$$R = \text{hydraulic ratio desired} = \frac{CaO}{SiO_2 + Fe_2O_3 + Al_2O_3}$$

Then

$$
\begin{aligned}
a &= S_1 - rO_1 \\
b &= rO_2 - S_2 \\
c &= rO_3 - S_3 \\
d &= L_1 - (S_1 + O_1)R \\
e &= (S_2 + O_2)R - L_2 \\
f &= (S_3 + O_3)R - L_3
\end{aligned}
$$

The proportions of the components is thus

Comp. 1	Comp. 2	Comp. 3
$ec - bf$	$ea - bd$	$cd - fa$

Or if Comp. 1 = 100, then

$$\text{Comp. } 2 = \frac{ea - bd}{ec - bf} \times 100$$

$$\text{Comp. } 3 = \frac{cd - fa}{ec - bf} \times 100$$

c) To obtain a given C_3S *and* C_3A *content*

1) Analysis of components (Raw Basis)

	L	*C*	*I*
SiO_2	.....	.....	......
Al_2O_3	.....	.....	.....
Fe_2O_3	.....	.....	.....
CaO	.....	.....	.....
Total	.....	.....	.....

2) Desired compound composition

Whatever targets are set, make sure to make:

$$C_3S + C_3A + (C_2S + C_4AF) + (MgO + \text{alkalies} + \text{loss}) = 100$$

$$C_3S = x_t = \ldots\ldots$$

$$C_3A = y_t = \ldots\ldots$$

$$C_2S + C_4AF = z_t = \ldots\ldots$$

3) Theoretical compound composition for each component

To be calculated from the raw analysis data. Also make sure to use the proper sign (+ or –).

		L	*C*	*I*
C_3S	:	$x_1 = \ldots\ldots$	$x_2 = \ldots\ldots$	$x_3 = \ldots\ldots$
C_3A	:	$y_1 = \ldots\ldots$	$y_2 = \ldots\ldots$	$y_3 = \ldots\ldots$
$C_2S + C_4AF$	:	$z_1 = \ldots\ldots$	$z_2 = \ldots\ldots$	$z_3 = \ldots\ldots$
Total	:	.....	.....	.....

Note: Proceed with the calculations only when the sum total of the compounds corresponds to the total of the oxides, i.e., for *L*: total oxides = $x_1 + y_1 + z_1$, etc.

4) Calculations, auxiliary matrix

$a = x_2/x_1 = \ldots\ldots$

$b = x_3/x_1 = \ldots\ldots$

$c = x_t/x_1 = \ldots\ldots$

$d = y_2 - y_1 a = \ldots\ldots$

$e = z_2 - z_1 a = \ldots\ldots$

$f = (y_3 - y_1 b)/d = \ldots\ldots$

$g = (y_t - y_1 c)/d = \ldots\ldots$

$h = z_3 - z_1 b - ef = \ldots\ldots$

$i = (z_t - z_1 c - eg)/h = \ldots\ldots$

5) Weight of each component needed per unit weight of clinker

Either English or metric weight units can be used, i.e., the results can be expressed in lb/lb, ton/ton, or kg/kg of clinker. Results obtained must all be positive numbers. If any of the results are negative, the desired mix cannot be obtained with the given raw materials. Either the targets have to be changed or other suitable raw materials must be selected.

for I = i = ...
for C = $g-fi$ = ...
for L = $c-aC-bi$ = ...
Total = W_u = weight of mix/wt. clinker = ...

6) Final Mix proportions

for I = I/W_u = ...
for C = C/W_u = ...
for L = L/W_u = ...

(Results are expressed as percent in form of decimals)

2.04 Four-Component Mix Calculation

a) Analysis of components (raw basis)

	L	C	I	S
SiO_2	.....	.....	.....	.....
Al_2O_3	.....	.....	.....	.....
Fe_2O_3	.....	.....	.....	.....
CaO	.....	.....	.....	.....

b) Desired clinker composition

(assume sum of MgO, total alkalies and ign. loss = 2 percent)

C_3S = x_t = ...
C_2S = y_t = ...
C_3A = z_t = ...
C_4AF = w_t = ...

c) Theoretical compound composition for each component

(to be calculated from raw analysis data, make sure to use the proper sign)

	L	*C*	*I*	*S*
C_3S :	x_1 = ...	x_2 = ...	x_3 = ...	x_4 = ...
C_2S :	y_1 = ...	y_2 = ...	y_3 = ...	y_4 = ...
C_3A :	z_1 = ...	z_2 = ...	z_3 = ...	z_4 = ...
C_4AF:	w_1 = ...	w_2 = ...	w_3 = ...	w_4 = ...

d) Raw material costs

(Insert here the total costs per ton, for each raw material. These costs will later be used to determine the cost of the calculated mix)

	L_c	C_c	I_c	S_c
# /ton:	.....	.....	.....	.....

Note: The sum total of the oxides of each raw material must equal the sum total of the compounds of that material.

e) Calculation, auxiliary matrix

$a = x_2/x_1$ = ...

$b = x_3/x_1$ = ...

$c = x_4/x_1$ = ...

$d = x_t/x_1$ = ...

$e = y_2 - y_1 a$ = ...

$f = z_2 - z_1 a$ = ...

$g = w_2 - w_1 a$ = ...

$h = (y_3 - y_1 b)/e$ = ...

$i = (y_4 - y_1 c)/e$ = ...

$j = (y_t - y_1 d)/e$ = ...

$k = z_3 - bz_1 - fh$ = ...

$l = w_3 - bw_1 - gh$ = ...

$m = (z_4 - z_1c - fi)/k$ = ...

$n = (z_t - z_1d - fj)/k$ = ...

$o = w_4 - w_1c - gi - lm$ = ...

$p = (w_t - w_1d - gj - ln)/o$ = ...

Note: Make sure to indicate the proper sign in the results.

f) Weight of each component required per unit weight of clinker

Either English or metric weight units can be used, i.e., the result can be expressed in terms of lb/lb clinker, ton/ton or kg/kg clinker. Results obtained must all be positive numbers.

for $S = p$ = ...
for $I = n - mp$ = ...
for $C = j - hI - ip$ = ...
for $L = d - aC - bI - cp$ = ...
Total = W_u = weight of mix/unit weight clinker. . . . .

g) Final mix proportions

for $S = S/W_u$ = ... percent

for $I = I/W_u$ = ... percent

for $C = C/W_u$ = ... percent

for $L = L/W_u$ = ... percent

Note: All results are expressed in terms of a decimal.

h) Cost of the mix

The cost of the mix per unit weight of clinker can be calculated as follows:

$$\#/\text{unit weight clinker} = [(\text{percent } S)(S_c)] + [(\text{percent } I)(I_c)] + [(\text{percent } C)(C_c)] + [(\text{percent } L)(L_c)] = \ldots\ldots$$

2.05 Determination of Chemical Composition

When certain properties are required in a mix, a preliminary investigation of the needed chemical composition can be made by the following trial and error method. This method is only applicable when $Al_2O_3/Fe_2O_3 = > 0.64$.

Desired:

a = SO_3 + MgO + Alkalies = ... percent
b = Fe_2O_3 = ... percent
c = C_3A = ... percent
d = SR (silica ratio) = ... percent
e = C_3S = ... percent

Step 1 $x_1 = 100 - a$ = ...

Step 2 $Al_2O_3 = (c + 1.692b)/2.65$ = ...

Step 3 $SiO_2 = d(Al_2O_3 + b)$ = ...

Step 4 $CaO = 0.24564(e + 7.6SiO_2 + 6.718Al_2O_3 + \ldots + 1.43b + 2.852SO_3)$ = ...

Step 5 $x = CaO + Al_2O_3 + b + SiO_2$ = ...

Step 6

x must equal $x_1 \pm 0.2$ percent to make the calculated chemical composition acceptable.

PROBLEMS AND SOLUTIONS

2.01 Given a clinker of 68.5 percent C_3S and a kilnfeed of 81.3 percent $CaCO_3$. What must the approximate $CaCO_3$ content be in the kiln feed to obtain a clinker of 61 percent C_3S?

$$CaCO_3 = 81.3 + 0.0806\,(61.0 - 68.5) = 80.70 \text{ percent} \quad (ans.)$$

2.02 (a) Given a limestone of $TC = 89.8$ and a clay of $TC = 21.0$. What percentage of each is needed to make a mix that contains 79.6 percent total carbonates?

$$79.6 = \frac{89.8x + (100 - x)21.0}{100}$$

$$68.8x = 5860$$

$$x = 85.17$$

Hence, the mix must contain 85.17 percent limestone and 100 – 85.17 = 14.83 percent clay. *(ans.)*

2.02 (b) How many kilograms of limestone are needed for every 100 kg of clay in the example given in **2.02 (a)**?

$$w = \frac{79.6 - 21.0}{89.8 - 79.6}\,100 = 574.5 \text{ kg limestone} \quad (ans.)$$

2.02 (c) Given:

Raw Basis

	SiO_2	Al_2O_3	Fe_2O_3	CaO	MgO
limestone	1.06	0.68	0.43	55.32	1.03
clay	50.10	18.60	21.80	2.10	0.01

What must the ratio be between these two raw materials in the mix?

$$x = 55.32 + [(1.4)(1.03)] - [(2.3)(1.06)] + [(1.7)(0.68)] + 0.43 = 52.738$$

$$y = [(2.3)(50.1)] + [(1.7)(18.6)] + 21.8 - 2.1 + [(1.4)(0.01)] = 166.536$$

Thus the required ratio is

$$\frac{166.536}{52.738} = 3.16:1 \quad (ans.)$$

2.02 (d) A kilnfeed tank contains 58 percent limestone and shows a $CaCO_3$ content of 78.4 percent. What percent limestone must the tank contain to obtain a final $CaCO_3$ content of 79.3 percent? The limestone used to make this correction shows 91.4 percent $CaCO_3$.

$$x = 58 + \frac{(79.3 - 78.4)(100 + 58)}{91.4 - 79.3} = 69.75 \text{ percent} \quad (ans.)$$

2.03 (a) Given:

Raw Basis

	Material I	*Material II*	*Material III*
SiO_2	1.06	68.18	9.98
Al_2O_3	0.68	11.94	1.66
Fe_2O_3	0.43	2.68	83.48
CaO	56.32	4.20	0.87

What percentage of each is needed to obtain a lime saturation factor of 0.935 and a silica ratio of 2.75 in the mix?

$a_1 = -52.272$ $a_2 = 1.9925$
$b_1 = 194.934$ $b_2 = -27.975$
$c_1 = -33.705$ $c_2 = -224.155$

for Material I, x = 41.57
for Material II, y = 10.97
for Material III, = 1.00
Total = 53.54

$$z = \frac{100}{53.54} = 1.8678$$

percent of Material I	$= 41.57z$	= 77.64	*(ans.)*
percent of Material II	$= 10.97z$	= 20.49	*(ans.)*
percent of Material III =	z	= 1.87	*(ans.)*
Total		= 100.0	

2.03 (c) Given the same three materials as in example **2.03 (a)**, determine the percentage of each needed to obtain C_3S = 62 percent, C_3A = 7.72 percent, and the sum $C_2S + C_4AF$ = 28.28 percent.

	L(I)	*C(II)*	*I(III)*	*Target*
C_3S	$x_1 =$ 216.04	$x_2 = -585.25$	$x_3 = -202.85$	$x_t = 62.0$
C_3A	$y_1 =$ 1.07	$y_2 =$ 27.11	$y_3 = -136.85$	$y_t =$ 7.72
$C_2S + C_4AF$	$z_1 = -158.63$	$z_2 =$ 645.14	$z_3 =$ 435.67	$z_t = 28.28$
Total :	58.48	87.00	95.97	98.00

$a = -2.7090$ $d = 30.0086$ $g = 0.2470$
$b = -0.9389$ $e = 215.4113$ $h = 1261.8777$
$c = 0.2870$ $f = -4.5269$ $i = 0.01633$

$I = 0.01633$ $= 0.01633$
$C = 0.2470 - [(4.5269)(0.01633)]$ $= 0.32092$
$L = 0.287 - [(-2.709)(0.32092)] - [(-0.9389)(0.01633)]$
$= 1.17170$
Total mix $= 1.5090$

Proportions:

$$I = \frac{0.01633}{1.5090} = 0.0108 = 1.08 \text{ percent} \quad (ans.)$$

$$C = \frac{0.32092}{1.5090} = 0.2127 = 21.27 \text{ percent} \quad (ans.)$$

$$L = \frac{1.1717}{1.5090} = 0.7765 = 77.65 \text{ percent} \quad (ans.)$$

Double check:

$$C_3S = [(1.1717)(216.04)] + [(0.32092)(-585.25)] + [(0.01633)(-202.85)] = 62.00$$

$$C_3A = [(1.1717)(1.07)] + [(0.32092)(27.11)] + [(0.01633)(-136.85)] = 7.72$$

$$C_2S + C_4AF = [(1.1717)(-158.63)] + [(0.32092)(645.14)] + [(0.01633)(435.67)] = 28.29$$

2.04 Given the four raw materials below, calculate the percentage of each needed to obtain a mix of 60.5 percent C_3S, 19.3 percent C_2S, 10.3 percent C_3A, and 8.6 percent C_4AF.

a) Analysis *(Raw basis)*

	L	*C*	*I*	*S*
SiO_2	7.64	68.15	50.21	48.91
Al_2O_3	0.49	13.68	35.95	21.43
Fe_2O_3	0.34	3.93	1.36	24.19
CaO	51.85	1.91	2.04	5.34
Total	60.32	87.67	89.56	99.87

b) Targets:

$x_t = 60.5$ $y_t = 19.3$ $z_t = 10.3$ $w_t = 8.6$

c) Compounds:

	L		C		I		S	
C_3S	$x_1 =$	149.22	$x_2 =$	−607.82	$x_3 =$	−616.85	$x_4 =$	−528.63
C_2S	$y_1 =$	−90.67	$y_2 =$	653.93	$y_3 =$	609.30	$y_4 =$	539.02
C_3A	$z_1 =$	0.72	$z_2 =$	29.60	$z_3 =$	92.97	$z_4 =$	15.86
C_4AF	$w_1 =$	1.03	$w_2 =$	11.96	$w_3 =$	4.14	$w_4 =$	73.61
Total		60.3		87.67		89.56		99.86

$a = -4.0733$ $\quad e = 284.6039$ $\quad i = 0.7653$ $\quad m = -0.0938$
$b = -4.1338$ $\quad f = 32.5328$ $\quad j = 0.1970$ $\quad n = 0.0521$
$c = -3.5426$ $\quad g = 16.1555$ $\quad k = 69.1426$ $\quad o = 64.4343$
$d = 0.4054$ $\quad h = 0.8239$ $\quad l = -4.9127$ $\quad p = 0.08157$

$S = \qquad = 0.08157$
$I = 0.0521 - [(-0.0938)(0.08157)] = 0.05975$
$C = 0.197 - [(0.8239)(0.05975)] - [(0.7653)(0.08157)] = 0.08535$
$L = 0.4054 - [(-4.0733)(0.08535)] - [(-4.1338)(0.05975)]$
$\qquad - [(-3.5426)(0.08157)] = 1.28902$
Total $= 1.51569$

$$\text{Percent } S = \frac{0.08157}{1.51569} = 0.0538 = 5.38 \text{ percent} \quad (\textit{ans.})$$

$$\text{Percent } I = \frac{0.05975}{1.51569} = 0.03942 = 3.94 \text{ percent} \quad (\textit{ans.})$$

$$\text{Percent } C = \frac{0.08535}{1.51569} = 0.05631 = 5.63 \text{ percent} \quad (\textit{ans.})$$

$$\text{Percent } L = \frac{1.28902}{1.51569} = 0.8505 = 85.05 \text{ percent} \quad (\textit{ans.})$$

Double check:

$$C_3S = [(1.28902)(149.22)] + [(0.08535)(-607.82)] + [(0.05975)(-616.85)] + [(0.08157)(-528.63)] = 60.49$$

(The other compounds, likewise, should be double checked in the same manner.)

2.05 A kilnfeed mix is desired that shows the following characteristics:

Fe_2O_3: 2.95, C_3A: 11.85, C_3S: 53.0 *S/R:* 2.40

What is the chemical composition of this mix if the sum total of the auxiliary oxides (MgO, SO_3, and alkalies) is expected to be 2.0 percent?

Solution:

$$x_1 = 100 - 2.0 = 98.0$$

$$Al_2O_3 = \frac{11.85 + [1.692(2.95)]}{2.65} = 6.36$$

$$SiO_2 = 2.40\,(6.36 + 2.95) = 22.34$$

$$CaO = 0.24564\left\{53.0 + [7.6(22.34)] + [6.718(6.36)] + [1.43\,(2.95)]\right\} = 66.26$$

$$x = 66.26 + 22.34 + 6.36 + 2.95 = 97.91$$

The sum total of the primary oxides is 97.91 and x_1 has been found earlier to be 98.0. Therefore, this composition is acceptable since the two agree closely with each other.

Chapter 3

KILN FEED SLURRY

3.01 Specific Gravity and Pulp Density of Slurries

a) English units

Table based on specific gravity = 2.7 for dry feed.

Percent slurry moisture	*Sp. gr. slurry*	*Pulp density* lb/ft^3	lb *dry solids per* ft^3
16	2.123	132.48	111.18
17	2.095	130.73	108.51
18	2.067	128.98	105.76
19	2.041	127.36	103.16
20	2.015	125.74	100.59
21	1.990	124.18	98.10
22	1.965	122.62	95.64
23	1.941	121.12	93.26
24	1.917	119.62	90.91
25	1.895	118.22	88.67
26	1.872	116.81	86.44
27	1.851	115.47	84.29
28	1.829	114.13	82.17
29	1.809	112.88	80.14
30	1.788	111.57	78.10
31	1.769	110.39	76.17
32	1.749	109.14	74.22
33	1.730	107.95	72.33
34	1.711	106.77	70.47
35	1.693	105.64	68.67
36	1.675	104.52	66.89
37	1.658	103.46	65.18
38	1.640	102.34	63.45
39	1.624	101.34	61.82
40	1.607	100.28	60.17
41	1.591	99.28	58.58
42	1.575	98.28	57.00
43	1.560	97.34	55.48
44	1.545	96.41	53.99
45	1.530	95.47	52.51
46	1.515	94.54	51.05
47	1.501	93.66	49.64
48	1.487	92.79	48.25
49	1.473	91.92	46.88
50	1.459	91.04	45.52

b) Metric units (SI)

Table is based on specific gravity of 2.7 for dry feed.

Percent slurry moisture	*Sp. gr. of slurry*	*Pulp density* kg/m³	kg *dry solids per* m³
16	2.123	2122.3	1781.8
17	2.095	2094.3	1738.3
18	2.067	2066.3	1694.3
19	2.041	2040.3	1652.6
20	2.015	2014.4	1611.5
21	1.990	1989.4	1571.6
22	1.965	1964.4	1532.2
23	1.941	1940.3	1494.0
24	1.917	1916.3	1456.4
25	1.895	1893.9	1420.5
26	1.872	1871.3	1384.8
27	1.851	1849.8	1350.3
28	1.829	1828.4	1316.4
29	1.809	1808.3	1283.8
30	1.788	1787.4	1251.2
31	1.769	1768.4	1220.2
32	1.749	1748.4	1189.0
33	1.730	1729.4	1158.7
34	1.711	1710.5	1128.9
35	1.693	1692.4	1100.1
36	1.675	1674.4	1071.6
37	1.658	1657.4	1044.2
38	1.640	1639.5	1016.5
39	1.624	1623.5	990.4
40	1.607	1606.5	963.9
41	1.591	1590.5	938.5
42	1.575	1574.4	913.1
43	1.560	1559.4	888.8
44	1.545	1544.5	864.9
45	1.530	1529.4	841.2
46	1.515	1514.5	817.8
47	1.501	1500.4	795.2
48	1.487	1486.5	773.0
49	1.473	1472.6	751.0
50	1.459	1458.5	729.2

3.02 Properties of Water

1 ft^3 of H_2O	=	7.4805 U.S. gal
	=	62.43 lb
1 gallon H_2O	=	8.342 lb
	=	0.1337 ft^3
1 m^3 H_2O	=	1000.00 liters
	=	1000.00 kg
1 liter H_2O	=	1000.00 cc (cm^3)
	=	1.00 kg
specific gravity of water	=	1.00 kgf/dm^3 @ 4°C
boiling point	=	100°C @ sea level
thermal conductivity	=	0.50 kcal/mh · C
specific heat	=	0.999 kcal/kg · C

3.03 Mass of Slurry Required per Mass of Clinker

$$w_1 = \frac{100F}{100-M}$$

English or metric units can be employed in this formula.

3.04 Slurry Feed Rate Required

English units

$$g = \frac{249.35cw_1}{d} = \frac{24{,}935Fc}{(100-M)d}$$

Metric units (SI)

$$G = \frac{Cw_1}{3.6D} = \frac{CF100}{3.6(100-M)D}$$

3.05 Clinker Production for a Given Slurry Rate

English units

$$c = \frac{gd}{249.35w_1} = \frac{(100-M)gd}{24{,}935F}$$

Metric units (SI)

$$C = \frac{3.6GD}{w_1} = \frac{3.6(100-M)DG}{100F}$$

C = clinker rate (long tons per hour)
c = clinker rate (short tons per hour)
w_1 = mass slurry per mass of clinker (tons/ton) or (kg/kg)
D = pulp density of slurry (kg/m^3)
d = pulp density of slurry (lb/ft^3)
F = mass dry feed per mass of clinker (tons/ton) or (kg/kg)
G = slurry rate (m^3/s)
g = slurry rate (gpm)
M = percent moisture

3.06 Clinker Production per Slurry Tank Unit

Note: This formula applies only to the cylindrical portion of the slurry tank.

English units

$$c_t = 0.08333\pi r^2 s/2000F$$

Metric units (SI)

$$C_T = \pi R^2 S/F$$

3.07 Specific Gravity of Slurry

$$s_{gw} = \frac{100}{M + (100 - M)(1/s_{gs})}$$

3.08 Dry Solids per Unit Volume of Slurry

English units

$$s = 62.43 s_{gw} [1 - (M/100)]$$

Metric units (SI)

$$S = 1000 s_{gw} [1 - (M/100)]$$

Note: values for S, s, s_{gw} and s_{gs} can also be obtained from **3.01**.

c_t = tons clinker per inch of slurry tank
C_T = kg clinker per meter of slurry tank height
F = mass of dry feed per mass of clinker (kg/kg) or (tons/ton)
R = radius of slurry tank (m)
r = radius of slurry tank (ft)
S = kg solids per m^3 of slurry
s = lb solids per ft^3 of slurry
s_{gs} = specific gravity of dry solids
s_{gw} = specific gravity of slurry

PROBLEMS AND SOLUTIONS

3.03 A given kiln uses a slurry of 32 percent moisture and the dry solids rate has been found to be 1.593 kg/kg clinker. What is the slurry consumption on this kiln?

$$w_1 = \frac{100(1.593)}{100 - 32} = 2.343 \text{ kg slurry/kg clinker} \quad (ans.)$$

3.04 In example **3.03** above, what is the slurry feed rate (lt/s) when the kiln produces 36,500 kg clinker/h?

Solution:

1. From Table 3.01 (b), find the pulp density of the slurry

$$D = 1748.4 \text{ kg/m}^3$$

2.

$$G = \frac{(36{,}500)(2.343)}{(3.6)(1748.4)} = 0.01359 \text{ m}^3\text{/s}$$

and :(1000)(0.01359) = 13.59 liters/s (*ans.*)

3.05 A kilnfeed tank has a diameter of 16.3 meters. The kiln receives slurry of 36 percent moisture and shows a specific dry feed consumption of 1.63 kg/kg clinker. How many kilograms of clinker are produced when the slurry level in the cylindrical portion of the tank drops 1 cm?

Solution:

From Table 3.01 (b), find the kg solids/m^3 of slurry.

$$S = 1071.6 \text{ kg/m}^3$$

$$c_T = \frac{(3.1416)(8.15)^2(1071.6)}{1.63} = 137{,}186 \text{ kg/m} =$$

$$1371.9 \text{ kg/cm} \quad (\textit{ans.})$$

3.07 Given: specific gravity of dry solids = 2.68. What is the specific gravity of the slurry when it contains 31.5 percent moisture?

$$s_{gw} = \frac{100}{31.5 + (100 - 31.5)(1/2.68)} = 1.753 \quad (\textit{ans.})$$

Chapter 4

CHEMICAL AND PHYSICAL PROPERTIES OF MATERIALS USED IN CEMENT MANUFACTURING

4.01 Chemical Compounds

Compound	Formula	Molecular Weight	Percent Composition	
Aluminum Oxide	Al_2O_3	102.2	0.5303 = Al	0.4697 = O
Aluminum Hydroxide	$Al_2(OH)_6$	156.26	0.654 = Al_2O_3	0.346 = H_2O
Calcium Carbonate	$CaCO_3$	100.07	0.5604 = CaO	0.4396 = CO_2
Calcium Chloride	$CaCl_2$	111.0	0.3613 = Ca	0.6387 = Cl
Calcium Fluoride	CaF_2	78.07	0.3613 = Ca	0.4866 = F
Calcium Hydroxide	$Ca(OH)_2$	74.11	0.7569 = CaO	0.2431 = H_2O
Calcium Oxide	CaO	56.07	0.7148 = Ca	0.2852 = O
Calcium Sulfate (anh.)	$CaSO_4$	136.13	0.412 = CaO	0.588 = SO_3
Calcium Sulfate (gypsum)	$CaSO_4 \cdot 2\,H_2O$	172.17	0.3258 = CaO 0.2092 = H_2O	0.465 = SO_3
Carbon Dioxide	CO_2	44.005	0.2727 = C	0.7273 = O
Carbon Monoxide	CO	28.005	0.4286 = C	0.5714 = O
Ethane	C_2H_6	30.08	0.7924 = C	0.2076 = H
Ferric Oxide	Fe_2O_3	159.68	0.6996 = Fe	0.3004 = O
Ferrous Oxide	FeO	71.84	0.7772 = Fe	0.2228 = O
Hydrogen Sulfide	H_2S	34.06	0.9407 = S	0.0593 = H

Compound	*Formula*	*Molecular Weight*	*Percent Composition*	
Magnesium Carbonate	$MgCO_3$	84.32	0.4784 = MgO	0.5216 = CO_2
Magnesium Oxide	MgO	40.32	0.6036 = Mg	0.3964 = O
Manganous Oxide	MnO	70.93	0.7746 = Mn	0.2254 = O
Manganic Oxide	Mn_2O_3	157.86	0.6962 = Mn	0.3038 = O
Manganese Dioxide	MnO_2	86.93	0.6322 = Mn	0.3678 = O
Methane	CH_4	16.04	0.7467 = C	0.2533 = H
Nitrous Oxide	N_2O	44.02	0.6365 = N	0.3635 = O
Nitric Oxide	NO	30.01	0.4668 = N	0.5332 = O
Nitrogen Peroxide	NO_2 or	46.01	0.3045 = N	0.6955 = O
	N_2O_4	96.02		
Phosphorous Pentoxide	P_2O_5	142.1	0.4366 = P	0.5634 = O
Potassium Oxide	K_2O	94.20	0.8303 = K	0.1697 = O
Potassium Sulfate	K_2SO_4	174.26	0.5408 = K_2O	0.4592 = SO_3
Silica	SiO_2	60.3	0.4702 = Si	0.5298 = O
Sodium Carbonate (anh.)	Na_2CO_3	106.0	0.5853 = Na_2O	0.4147 = CO_2
Sulfur Dioxide	SO_2	64.06	0.5005 = S	0.4995 = O
Sulfur Trioxide	SO_3	80.06	0.4005 = S	0.5995 = O
Titanium Dioxide	TiO_2	80.01	0.6004 = Ti	0.3996 = O
Zinc Oxide	ZnO	81.37	0.8034 = Zn	0.1966 = O

4.02 Bulk Densities of Common Materials

	lb/ft^3	kg/m^3
Aluminum	162	2595
Asbestos	190	3045
Brick (basic)	150–185	2400–2965
(alu.)	95–110	1520–1760
(fireclay)	85–95	1360–1520
Cement (packed)	94	1506
(loose)	75–90	1200–1440
Clay (loose)	60–75	960–1200
Clinker	90–106	1440–1700
Coal (loose)	50–54	800–865
Coke	30–40	480–640
Concrete (reinforced)	145	2325
Gravel (loose)	110	1760
Ice	57.4	919
Iron (Cast)	450	7210
Iron Ore	175	2805
Kiln Feed (dry)	85	1360
Kiln Dust (loose)	65	1040
Limestone	95	1520
Mortar	104	1665
Fuel Oil	56	895
Sand	95	1520
Shale	155	2480
Slurry (@ 35 percent H_2O)	105	1682
Steel	490	7850
Water	62.4	1000

4.03 Typical Coal Analysis

(As received Basis)

Type of Coal	*Moisture*	*Volatile Matter*	*Free Carbon*	*Ash*	S	H	C	N	O	Btu/lb (*gross*)	kcal/kg	J/g
Lignite	33.4	40.4	17.2	9.0	.6	3.1	40.8	.8	12.3	7500	4167	17400
Sub-Bituminous	22.3	31.4	34.7	11.6	2.6	3.2	70.3	1.0	11.3	8300	4610	19300
Bituminous high volatile	12.0	34.2	47.4	9.3	.5	4.4	73.4	1.3	11.1	11500	6390	26750
Bituminous low volatile	3.6	15.4	76.3	11.7	.8	4.6	79.0	1.4	2.5	13000	7220	30240
Anthracite	5.4	7.0	71.8	15.8	.8	2.5	77.9	.8	2.2	12000	6670	27900

4.04 Typical Fuel Oil Properties

Grade	*Type*	*API gravity 60*	*Specific gravity 60/60*	lb/ gal	kg/ m^3	*visc. centi- stokes* 100°F	S	O_2 + N_2	H_2	C	Btu/ gal *(in thousands)*	kcal/ kg *(in thousands)*	J/g *(in thousands)*
No. 1	Kerosene	40	0.8251	6.87	823	1.6	.1	.2	13.2	86.5	137	35.9	46.4
No. 2	Light Oil	32	0.8654	7.21	864	2.7	.7	.2	12.7	86.4	141	35.2	45.5
No. 6	Bunker "C"	12	0.9861	8.21	984	360.0	2.5	.9	10.4	85.3	148	32.5	42.0

4.05 Typical Gaseous Fuel Properties

Natural Gas

Methane CH_4	*Ethane* C_2H_6	*Propane* C_3H_8	*Butane* C_4H_{10}	*Pentane* C_5H_{12}	CO_2	N_2	H_2S	lb/ft^3	kg/m^3	Btu/ft^3 *(gross)*	kcal/m^3
77.73	5.56	2.4	1.18	0.63	5.5	–	7.0	0.0562	0.9005	1061	9442

4.06 Barometric Pressure at Different Altitudes

Elevation ft	m	in. Hg	p.s.i.	mm Hg	*kPa*
0	0	29.90	14.68	760	101.22
100	30	29.79	14.63	757	100.87
200	60	29.68	14.57	754	100.46
300	90	29.57	14.52	752	100.11
400	121	29.46	14.46	749	99.70
500	152	29.36	14.42	746	99.42
600	183	29.25	14.36	743	99.01
700	213	29.14	14.31	741	98.66
800	244	29.03	14.25	738	98.25
900	274	28.92	14.20	735	97.91
1000	305	28.82	14.15	733	97.56
1200	366	28.61	14.05	727	96.87
1400	427	28.40	13.94	722	96.11
1600	488	28.19	13.84	717	95.42
1800	549	27.98	13.74	711	94.73
2000	610	27.78	13.64	706	94.04
2200	671	27.57	13.54	701	93.36
2400	732	27.27	13.44	696	92.67
2600	792	27.17	13.34	691	91.98
2800	853	26.97	13.24	686	91.29
3000	914	26.77	13.14	680	90.60
3200	975	26.58	13.05	676	89.98
3400	1036	26.38	12.95	671	89.29
3600	1097	26.19	12.86	666	88.67
3800	1158	26.00	12.77	661	88.05
4000	1219	25.81	12.67	656	87.36
4500	1372	25.34	12.44	644	85.77
5000	1524	24.88	12.22	632	84.25
5500	1676	24.43	11.99	621	82.67
6000	1829	23.98	11.77	610	81.15
6500	1981	23.54	11.56	598	79.70

4.07 Sieve Sizes

DIN 4188			ASTM E-11-60			
mesh per cm^2	Aperture mm	Aperture microns	mesh per inch	Aperture mm	Aperture in.	Aperture microns
			325	0.044		44
6400	0.075	75	200	0.074		74
4900	0.090	90	170	0.088		88
3600	0.1	100	150	0.105		105
1600	0.150	150	100	0.149		149
400	0.3	300	48	0.297	0.0117	297
144	0.5	500	32	0.5	0.0197	500
100	0.6	600	28	0.59	0.0232	

4.08 Coefficients of Linear Expansion

Substance	*Coefficient of linear expansion* in./in./°F	m/m/°C
Aluminum	0.00001244	22.70
Brick	0.0000035	6.41
Concrete	0.0000080	14.64
Copper	0.0000090	16.50
Cast Iron	0.00000655	11.99
Wrought Iron	0.00000661	12.09
Steel (mild)	0.00000636	11.64

4.09 Properties of Air

a) English units

°F	*Volume of dry air* @ 29.9 in. Hg ft^3/lb	*Volume of saturated air* ft^3/lb	*Weight of dry air* @ 29.9 in. Hg lb/ft^3	*Weight of water required to saturate air* lb *vapor*/lb *air*
32	12.39	12.46	0.080714	0.003774
60	13.09	13.33	0.076366	0.01104
70	13.35	13.68	0.074924	0.01576
90	13.85	14.54	0.072198	0.03106

b) Metric units (SI)

°C	*Volume of dry air* @ 1 atm. m^3/kg	*Volume of saturated air* m^3/kg	*Mass of dry air* @ 1 atm. kg/m^3	*Mass of water required to saturate air* kg *vapor*/kg *air*
0	0.7735	0.7779	1.2929	0.003774
16	0.8175	0.8322	1.2232	0.01104
21	0.8332	0.8540	1.2001	0.01576
32	0.8647	0.9077	1.1565	0.03106

4.10 Particulate Concentration in Gases

For gases:

1 lb/ft^3 = $385.1 \times 10^6\, m$ ppm
1 μg/liter = $24.04\, m$ ppm
1μg/m^3 = $0.02404\, m$ ppm

where

m = molecular weight of gas

Particulate:

1 grain/ft^3 = 2.29×10^6 μg/m^3
= 1.43×10^{-4} lb/ft^3

1 lb/ft^3 = 16.02×10^6 mg/m^3
= 16.02 grams/liter
= 7.06×10^3 grains/ft^3

1 mg/m^3 = 28.32 μg/ft^3
= 62.43×10^{-9} lb/ft^3

1 μg/m^3 = 0.001 mg/m^3
= 4.37×10^{-7} g/ft^3

To calculate the particulate concentration in a known volume of gas, use the following formula:

$$c = (2.205 \times 10^{-6})(m/V)$$

where

c = concentration, lb/ft^3
m = total particulate matter, mg
V = volume of gas sample through dry gas meter (standard condition), ft^3.

4.11 Selected International Atomic Weights

(Based on Carbon–12)

Element	*Symbol*	*Atomic Number*	*Atomic Weight*
Aluminum	Al	13	26.9815
Calcium	Ca	20	40.08
Carbon	C	6	12.0115
Chlorine	Cl	17	35.453
Fluorine	F	9	18.9984
Hydrogen	H	1	1.00797
Iron	Fe	26	55.847
Lead	Pb	82	207.19
Magnesium	Mg	12	24.312
Manganese	Mn	25	54.938
Nitrogen	N	7	14.0067
Oxygen	O	8	15.9994
Phosphorus	P	15	30.9738
Potassium	K	19	39.102
Silicon	Si	14	28.086
Sodium	Na	11	22.9898
Sulfur	S	16	32.064
Titanium	Ti	22	47.90
Zinc	Zn	30	65.37

4.12 Selected Minerals and Ores

Principal mineral or element	*Found in*	*Other nomenclature*
Aluminum	Bauxide	Hydrous aluminum oxide
	Gibbsite	
	Boehmite	
	Diaspore	
	Cryolite	
Aluminum silicates (clays)	Montmorillonite	
	Andalusite	
	Kyanite	
	Sillimanite	
	Dumortierite	Aluminum borosilicate
	Keolinite	
Asbestos	Tremolite	Calcium magnesium amphibole
	Actinolite	Calcium magnesium-iron amphibole
	Crysotile	Fibrous serpentine
Barium	Barite	Barium sulfate
	Witherite	Barium carbonate
Bentonite	Montmorillonite	
Borates	Borax	
	Ulexite	
	Colemanite	
	Kernite	
	Probertite	
Cadmium	Greenockite	Cadmium sulfide
Chromite	Chromite	Chrom-iron oxide
Cobalt	Cobaltite	Sulfarsenide of cobalt
	Smaltite	Cobalt diarsenide
	Erithrite	Cobalt bloom

Principal mineral or element	*Found in*	*Other nomenclature*
Copper	Chalcopyrite	Copper-iron sulfide
	Bornite	
	Cupriferrous pyr.	
	Covellite	
	Tetrahedrite	
	Enarcite	
	Cuprite	
	Atacamite	
	Malachite	
	Azurite	
	Chalcanthite	
	Crysocolla	
Diatomite	Diatomaceous earth	
Feldspars	Sanadine	Orthoclase of potassium or sodium
	Adularia	Orthoclase
	Albite	Sodium & aluminum silicate
	Anorthite	Calcium-aluminum silicate
Fluorspar	Fluorite	Calcium fluoride
Gypsum	Gypsum	Hydrous calcium sulfate
Iron	Hematite	Fossil ore (red)
	Limonite	Bog ore (brown)
	Magnetite	Lodestone (black)
	Siderite	Black band, kidney ore
Limestone	Calcite	Calcium carbonate
	Dolomite	Calcium-magnesium carbonate
	Alkerite	Calcium-magnesium-iron carbonate
	Aragonite	Orthorhombic calcium carbonate
	Manganocalcite	
Magnesium	Magnesite	Magnesium carbonate
	Dolomite	Pearlspar
	Brucite	Magnesium hydroxide
	Carnallite	Hydrous potassium-magnesium chloride

Principal mineral or element	*Found in*	*Other nomenclature*
Manganese	Psilomelane	Manganese oxide
	Pyrolusite	Manganese dioxide
Phosphates	Collophane	Hydrous calcium phosphate
	Apatite	Calcium-chloro-fluoro phosphate
	Wavellite	Hydrous alumina phosphate
	Voelckerite	
Pyrite & Sulfur	Pyrite	Iron disulfide
Quartz	Sand	Silicon dioxide
	Quartzite	
	Amethyst	Transparent quartz
	Smoky quartz	Smoky yellow, black
	Citrine	
	Rose quartz	
	Aventurine	
	Chalcedony	
	Chrysoprase	
	Heliotrope	
	Agate	
Titanium	Ilmenite	Ferrous titanate
	Rutile	Titanium oxide
	Sphene	Calcium-titano silicate

4.13 Classification of Minerals

a) Igneous rock

These are formed by the intrusion or extrusion of magma or by volcanic activity. The following minerals belong to this group:

Granite	:	Crystalline quartz and orthoclase.
Orthoclase	:	Feldspar containing potassium.
Plagioclase	:	Feldspar containing calcium and sodium.
Quartz	:	Silicon dioxide
Biotite	:	A dark green form of mica consisting of silicate of Fe, Mg, K, or Al.
Pyroxene	:	Silicate minerals containing calcium, sodium, magnesium, iron, or aluminum.
Olivine	:	Silicate of magnesium and iron.
Magnetite	:	Oxide of iron.

b) Sedimentary rock

These are formed by deposits of sedimentation. They can also consist of fragments of rock deposited in water or by precipitation from solutions and organisms. The following rocks belong to this group:

Gravel	Shale	Chalk
Sandstone	Limestone	Marl
Siltstone	Gypsum	Coral

c) Metamorphic rock

These are minerals that have been changed by the action of heat, pressure, and water. The following minerals belong to this group:

Gneiss	:	Laminated or foliated metamorphic rock
Schist	:	Crystalline metamorphic rock with foliated structure along parallel planes.
Slate	:	Compressed clays and shales.
Marble	:	Metamorphic crystallized limestone.
Quartzite	:	Metamorphic sandstone

4.14 Chemical Formula and Molecular Weight of Common Minerals

Name	*Synonym*	*Formula*	*Mol. wt.*
Aluminum hydroxide	Diaspore, boehmite	$AlO(OH)$	59.99
Aluminum hydroxide	Gibbsite	$Al(OH)_3$	78.00
Aluminum oxide	Corundum	Al_2O_3	101.96
Aluminum oxide	Trihydrate of aluminum	$Al_2O_3 \cdot 3H_2O$	156.01
Aluminum silicate	Andalusite, sillimanite	$Al_2O_3 \cdot SiO_2$	162.04
Aluminum silicate	Mullite	$3Al_2O_3 \cdot 2SiO_2$	426.05
Calcium aluminate		$CaAl_2O_4$	158.04
Tricalcium aluminate	C3A	$Ca_3Al_2O_3$	270.20
Calcium carbonate	Calcite, aragonite	$CaCO_3$	100.09
Calcium chloride		$CaCl_2$	110.99
Calcium fluoride	Fluorite	CaF_2	78.08
Calcium hydroxide		$Ca(OH)_2$	74.09
Calcium magnesium carbonate	Dolomite	$CaCO_3MgCO_3$	184.41
Calcium oxide		CaO	56.08
Tricalcium silicate	Alite, C3S	Ca_3SiO_5	228.32
Dicalcium silicate	Belite, C2S	$2CaO \cdot SiO_2$	172.24
Calcium sulfate	Anhydrite	$CaSO_4$	136.14
Calcium sulfate, hemihydrate	Plaster of Paris	$CaSO_2 \cdot \frac{1}{2}H_2O$	145.15
Calcium sulfate, dihydrate	Gypsum	$CaSO_4 \cdot 2H_2O$	172.17
Calcium sulfide	Oldhamite	CaS	72.14

Name	*Synonym*	*Formula*	*Mol. Wt.*
Calcium sulfite		$CaSO_3 \cdot 2H_2O$	156.17
Copper carbonate	Malachite	$CuCO_3Cu(OH)_2$	221.11
Copper oxide	Chalcocite, cuprite	Cu_2O	143.08
Iron carbonate	Siderite	$FeCO_3$	115.85
Iron disulfide	Pyrite, marcasite	FeS_2	119.98
Iron titaniumate	Ilmenite	$FeTiO_3$	151.75
Iron oxide	Wuestite	FeO	71.85
Iron oxide	Hematite	Fe_2O_3	159.69
Iron oxide	Magnetite	Fe_3O_4	231.54
Magnesium aluminate	Spinel	$MgAl_2O_4$	142.27
Magnesium carbonate	Magnesite	$MgCO_3$	84.32
Magnesium hydroxide	Brucite	$Mg(OH)_2$	58.33
Magnesium oxide	Periclase	MgO	40.31
Magnesium silicate	Enstatite	$MgSiO_3$	100.39
Magnesium orthosilicate	Forsterite	Mg_2SiO_4	140.71
Magnesium sulfate		$MgSO_4$	120.37
Manganese oxide	Hausmannite	Mn_3O_4	228.81
Manganese dioxide	Pyrolusite	MnO_2	86.94
Manganese sesquioxide	Braunite	Mn_2O_3	157.87
Manganese hydroxite	Manganite	$MnO(OH)$	87.94
Manganese oxide	Manganosite	MnO	70.94
Potassium aluminosilicate	Orthoclase	$KAlSi_3O_8$	278.34
Potassium carbonate		K_2CO_3	138.21
Potassium chlorate		$KClO_3$	122.55

Name	*Synonym*	*Formula*	*Mol. Wt.*
Potassium hydroxide		KOH	56.11
Potassium nitrate	Saltpeter	KNO_3	101.11
Potassium metasilicate		K_2SiO_3	154.29
Potassium sulfate	Arcanite	K_2SO_4	174.27
Silicon dioxide	Quartz, christobalite	SiO_2	60.08
Sodium alumina trisilicate	Albite	$NaAlSi_3O_8$	262.22
Sodium carbonate		Na_2CO_3	105.99
Sodium chloride	Common salt, halite	NaCl	58.44
Sodium alumina ferrite	Cryolite	Na_3AlFe_6	429.43
Sodium monoxide		Na_2O	61.98
Sodium silicate	Waterglass	$Na_2O \cdot xSiO_2$	
Sodium sulfate		Na_2SO_4	142.04
Stannous dioxide	Cassiterite	SnO_2	150.69
Titanium dioxide	Rutile, anatase, brookite	TiO_2	79.90
Titanium sesquioxide		Ti_2O_3	143.80
Zinc carbonate	Smithonite	$ZnCO_3$	125.39
Zinc oxide	Zincite	ZnO	81.37

Chapter 5

FORMULAS AND DATA USED IN COMBUSTION CALCULATIONS

5.01 Thermochemical Reactions

Formula:	C	+	O_2	=	CO_2
Molecular weight:	12	+	32	=	44
Mass:	1	+	2.66	=	3.66 lb + 14,093 Btu/lb
					3.66 kg + 7829 kcal/kg
	C	+	½ O_2	=	CO
	12	+	16	=	28
	1	+	1.33	=	2.33 lb + 4320 Btu/lb
					2.33 kg + 2400 kcal/kg
	2 H_2	+	O_2	=	2 H_2O
	4	+	32	=	36
	1	+	8	=	9 lb + 60,991 Btu/lb (gross)
					9 kg + 28,641 kcal/kg (H_u)
	S	+	O_2	=	SO_2
	32	+	32	=	64
	1	+	1	=	2 lb + 3983 Btu/lb
					2 kg + 2213 kcal/kg

and

SO_2	+	½ O_2	=	SO_3
64	+	16	=	80
1	+	0.25	=	1.25

5.02 Combustion Constants

a) English units

	Formula	lb/ft³	Btu/ft³	Btu/lb
Carbon	C			14,093
Hydrogen	H_2	0.005327	324.9	60,991
Oxygen	O_2	0.08461		
Nitrogen	N_2	0.07439		
Sulfur	S			3,983
Carbon monoxide	CO	0.07404	321.6	
Carbon dioxide	CO_2	0.117		
Water vapor	H_2O	0.04758		
Sulfur dioxide	SO_2	0.1733		
Methane	CH_4	0.04246	1014.6	23,896
Ethane	C_2H_6	0.08029	1789.0	22,282
Propane	C_3H_8	0.1196	2573.0	21,523
n-Butane	C_4H_{10}	0.1582	3392.0	21,441
n-Pentane	C_5H_{12}	0.1904	4200.0	22,058

Note: Volumes at 60°F and 30 in. Hg. All gross heating values.

b) Metric units

	Formula	kg/m³	kcal/m³	MJ/m³	kcal/kg	J/g
Carbon	C	–	–		7,831	32,765
Hydrogen	H_2	0.0853	2,892	12,098	33,884	141,771
Oxygen	O_2	1.3553	–	–	–	–
Nitrogen	N_2	1.1916	–	–	–	–
Sulfur	S	–	–	–	2,213	9,259
Carbon monoxide	CO	1.1860	2,862	11.975	–	–
Carbon dioxide	CO_2	1.8742	–	–	–	–
Water vapor	H_2O	0.7622	–	–	–	–
Sulfur dioxide	SO_2	2.7760	–	–	–	–
Methane	CH_4	0.6801	9,030	37.781	13,276	55,547
Ethane	C_2H_6	1.2861	15,922	66.618	12,379	51,794
Propane	C_3H_8	1.9158	22,900	95.812	11,957	50,028
n-Butane	C_4H_{10}	2.5341	30,189	126.309	11,912	49,840
n-Pentane	C_5H_{12}	3.0499	37,380	156.398	12,254	51,271

Note: Volumes at 16°C and 760 mm Hg.

5.03 Heat Value of Fuel

The heat value of a fuel is usually determined in a calorimeter. For an approximate indication, the heat value can also be calculated from the ultimate analysis. Values for C (carbon), S (sulfur), H (hydrogen), etc., are expressed in terms of percent by weight for coal and oil and in terms of percent by volume for natural gas.

For coal:

$$\text{gross heating value (Btu/lb)} = 145.4\,C + 40.4\,S + 611.0\,H - 64.6\,O_2$$

$$\text{net heating value (Btu/lb)} = 145.4\,C + 40.4\,S + 516.6\ H - \frac{O_2}{8} - 10.8m$$

$$\text{net heating value (kcal/kg)} = 80.8\,C + 287\left(H - \frac{O_2}{8}\right) + 22.45\,S - 6.0m$$

where

m = percent moisture in coal.

For oil:

$$\text{gross heating value (Btu/lb)} = 17{,}780 + 54\,(\text{API gravity})$$

$$\text{heating value (kcal/kg)} = 12{,}958 - 3228d - 70.0\,S$$

where

d = density at 15°C (kg/dm^3)

For natural gas:

$$Btu/ft^3 = 10.146\ CH_4 + 17.89\ C_2H_6 + 25.73\ C_3H_8 + 33.92\ C_4H_{10} + 42.0\ C_5H_{12} + 6.47\ H_2S$$

$$kcal/m^3 = 90.3\ CH_4 + 159.2\ C_2H_6 + 229 C_3H_8 + 301.9\ C_4H_{10} + 373.8\ C_5H_{12} + 57.6\ H_2S$$

5.04 Conversion from "Gross" to "Net" Heating Value

The net heating value accounts for the heat losses that are experienced for the evaporation of the moisture in the fuel as well as the water that is generated by the combustion of hydrogen. Heating values obtained in the calorimeter are "gross" values and can be converted to the "net" basis by the following formulas:

a) English units

$$HV_{(net)} = HV_{(gross)} - 9270\ H_2 \qquad (Btu/lb)$$

b) Metric units

$$Hu_{(net)} = Hu_{(gross)} - 5150\ H_2 \qquad (kcal/kg)$$

where

H_2 = percent hydrogen (sum total of H_2 in the fuel and the moisture)

In Europe it is the custom to express the heating value or fuel consumptions in terms of the "net" basis whereas in North America the "gross" heating value is generally used.

5.05 Analysis of Coal

a) Ultimate analysis

$$C + H + N + S + O + Ash = 100 \text{ percent (by weight)}$$

where

C = percent carbon, H = percent hydrogen, N = percent nitrogen,
S = percent sulfur, O = percent oxygen.

The percent oxygen (O) is not determined by analytical methods but calculated by difference to make the sum total equal to 100.

b) Proximate analysis

$$V + \text{free C} + \text{ash} + m = 100 \text{ percent}$$

where

V = percent volatiles,
free C = percent free carbon, and
m = percent moisture.

The percent free carbon is calculated by difference to make the sum total equal to 100.

5.06 Methods of Expressing Solid Fuel Analysis

Analysis of solid fuels should be reported with a note containing a remark in respect to the method in which the analysis is expressed. The following are the methods (basis) frequently used:

a) "as analyzed"
b) "dry basis"
c) "as received"
d) "combustible basis" (moisture and ash free)
e) "as fired"

For inventory control purposes it is of advantage to express coal tonnage, heating value and its composition on the "dry basis" to eliminate the fluctuations coal undergoes when it is stored outside and exposed to weathering.

5.07 Conversion of Coal Analysis to Different Basis

Let

Y = percent C, S, N, or percent ash
O = percent oxygen
H = percent hydrogen
m = percent moisture

subscript:

a = "as analyzed" basis
d = "dry basis"
r = "as received" basis
f = "as fired" basis
c = "combustible" basis

a) To convert from "as analyzed" to "dry" basis

$$Y_d = Y_a \frac{100}{100 - m_a}$$

$$H_d = (H_a - 0.1119 m_a) \frac{100}{100 - m_a}$$

O_d = calculated by difference

b) To convert from "dry" to "as received" basis

Multiply all components, except hydrogen, by the factor

$$\frac{100 - m_r}{100}$$

c) To convert from "dry" to "as fired" basis

Multiply all components, except hydrogen, by the factor

$$\frac{100 - m_f}{100}$$

Note: in b) and c) above, the percent hydrogen is calculated as follows:

$$H = H_d \frac{100 - m}{100} + 0.1119m$$

d) To convert from "as received" to "dry" basis

Multiply each component, except the hydrogen, by the factor

$$\frac{100}{100 - m_r}$$

$$H_d = (H_r - 0.1119m_r) \frac{100}{100 - m_r}$$

O_d = calculated by difference

e) To convert from "combustible" to "as fired" basis

Multiply each component, except the hydrogen, by the factor

$$\frac{100 - (\text{ash} + m_f)}{100}$$

$$H_f = H_c \frac{100 - (\text{ash} + m_c)}{100} + 0.1119 m_c$$

$$O_f = \text{calculated by difference}$$

f) To convert from "as received" to "combustible" basis

Multiply each component, except the hydrogen, by the factor

$$\frac{100}{100 - (\text{ash} + m_r)}$$

$$H_c = (H_r - 0.1119 m_r) \frac{100}{100 - (\text{ash} + m_r)}$$

The following table shows clearly how the values of a coal analysis and the heating value can change when the analysis is expressed in different terms.

	As analyzed	*Dry basis*	*Ash and moisture free*	*As received*	*As fired*
C (carbon)	61.24	67.30	75.36	57.00	66.76
H (hydrogen)	5.74	5.20	5.93	6.12	5.25
S (sulfur)	2.46	2.70	3.03	2.29	2.68
N (nitrogen)	1.73	1.90	2.12	1.61	1.88
O (oxygen)	19.09	12.20	13.56	8.62	12.02
Ash	9.74	10.70	–	9.06	10.61
m (moisture)	9.0	–	–	15.3	0.80
Volatiles	32.49	35.70	39.97	30.24	35.41
Free Carbon	48.78	53.60	60.02	45.40	53.17
Btu/lb	10,785	11,850	13,320	10,040	11,800
kcal/kg	5,992	6,583	7,400	5,578	6,556
kJ/kg	25,071	27,543	30,962	23,338	27,430

5.08 Typical Coal Ash Analysis

For a cement chemist, it is important to know the chemical composition of the coal ash. The majority of the ash, during the burning of coal, enters the clinker and modifies its chemical composition. On coal fired kilns, it is not only important to maintain a uniform kiln feed but also to fire the kiln with a coal of uniform composition. In plants, where coal originates from several different suppliers, provisions should be made to blend these coals before they are fired in the kiln. A typical analysis of coal ash is shown in the following:

SiO_2	Al_2O_3	Fe_2O_3	CaO
46.3 percent	25.6 percent	18.7 percent	2.8 percent

5.09 Fuel Ignition Temperatures

The approximate ignition temperatures of various fuels are

	°F	°C
Coal	480	250
Wood	570	300
Bunker C oil	400	200
Diesel fuel	650	350
Natural gas	1050	550

5.10 Percent Coal Ash Absorbed in Clinker

The percent coal ash contained in the clinker can be calculated from the loss-free analysis of the ash, raw mix, and clinker as follows:

Analysis

	CaO	SiO_2	Al_2O_3	Fe_2O_3
Ash	C_a =.....	S_a =.....	A_a =.....	F_a =.....
Raw Mix	C_m =.....	S_m =.....	A_m =.....	F_m =.....
Clinker	C =.....	S =.....	A =.....	F =.....

$$x_1 = \frac{C - C_m}{C_a - C_m} 100 = \ldots\ldots$$

$$x_2 = \frac{S - S_m}{S_a - S_m} 100 = \ldots\ldots$$

$$x_3 = \frac{A - A_m}{A_a - A_m} 100 = \ldots\ldots$$

$$x_4 = \frac{F - F_m}{F_a - F_m} 100 = \ldots\ldots$$

The percent coal ash contained in the clinker (percent) is:

$$\text{Percent} = \frac{x_1 + x_2 + x_3 + x_4}{4} = \ldots\ldots$$

5.11 Effect of Coal Ash on Clinker Composition

Changes in the composition of the clinker as a result of coal ash addition can be calculated by the following method:

Analysis (loss-free)

	CaO	SiO_2	Al_2O_3	Fe_2O_3	MgO	*Alkalies*	*Total*
Ash	C_a	S_a	A_a	F_a	M_a	N_a	
Clinker	C	S	A	F	M	N	X ....

w = coal factor (weight coal/wt. cl.), decimal = ...
y = ash content of coal, decimal = ...
z = ash absorption by clinker (usually 0.40 to 1.00) = ...

$\Delta C = wyC_a z =$ ..... $\Delta F = wyF_a z =$ .....
$\Delta S = wyS_a z =$ ..... $\Delta M = wyM_a z =$ .....
$\Delta A = wyA_a z =$ ..... $\Delta N = wyN_a z =$ .....

			clinker composition
CaO	$= C - \Delta C$	= (....)(X/v)	= .....
SiO_2	$= S - \Delta S$	= (....)(X/v)	= .....
Al_2O_3	$= A - \Delta A$	= (....)(X/v)	= .....
Fe_2O_3	$= F - \Delta F$	= (....)(X/v)	= .....
MgO	$= M - \Delta M$	= (....)(X/v)	= .....
Alkalies	$= N - \Delta N$	= (....)(X/v)	= .....
v	= Total	=	

Note: for (w) above either the English (lb coal/lb clinker) or metric units (kg coal/kg clinker) can be used.

5.12 Determination of Theoretical Fuel Consumption

Knowing the properties of the coal, kiln feed, and the exit gas allows an engineer to calculate the coal consumption by using Dr. Kuhl's formuıa:

Data needed:

a = constant, 0.266
b = percent carbon in dry coal
c = percent hydrogen in dry coal
d = percent nitrogen in dry coal
e = percent oxygen in dry coal
f = percent sulfur in dry coal
g = percent ash in dry coal

m = percent total carbonates in dry feed ($CaCO_3$ + $MgCO_3$)
n = percent loss on ignition of dry feed
p = percent "true" CO_2 in exit gas. Calculated as follows:

$$p = \frac{100\ CO_2 + CO}{100 + 1.89\ CO - 4.78\ O_2}$$

q = percent O_2 (oxygen) in exit gas
s = percent N_2 (nitrogen) in exit gas
t = percent CO (carbon monoxide) in exit gas

Calculations

$$K_1 = \frac{400m}{100 - n}$$

$$K_2 = \frac{b}{3} + c + a\frac{d}{7} - \frac{e}{8} + \frac{f}{8}$$

$$K_3 = \frac{b}{3} - \frac{mg}{2500 - 25n}$$

$$V = as - q + \frac{t}{2}$$

$$W = p + t$$

coal consumption,

$$x = \frac{K_1 V}{K_2 W - K_3 V}$$

where

x = lb coal/100 lb clinker or kg coal/100 kg clinker

Example: A kiln under investigation showed the following analysis during the time of the test:

Coal (dry)	
percent carbon	: 72.16
percent hydrogen	: 5.37
percent nitrogen	: 1.58
percent oxygen	: 7.89
percent sulfur	: 2.00
percent ash	: 11.00

Kiln feed (dry)	
total carbonates	: 79.26
loss on ignition	: 35.8

Exit gas	
CO_2 (true)	: 27.38
O_2	: 1.6
CO	: —
N_2	: 70.99

What is the coal consumption on this kiln?

$$K_1 = \frac{(400)(79.26)}{100 - 35.8} = 493.83$$

$$K_2 = \frac{72.16}{3} + 5.37 + 0.266\ \frac{1.58}{7} - \frac{7.89}{8} + \frac{2.00}{8} = 28.75$$

$$K_3 = \frac{72.16}{3} - \frac{(79.26)(11.00)}{2500 - [(25)(35.8)]} = 23.51$$

$$V = (0.266)(70.99) - 1.6 + \frac{0}{2} = 17.28$$

$$W = 27.38 + 0 = 27.38$$

$$x = \frac{(493.83)(17.28)}{(28.75)(27.38) - [(23.51)(17.28)]}$$

$$= 22.4 \text{ lb coal/100 lb clinker}$$

or

$$= 22.4 \text{ kg coal/100 kg clinker} \qquad (\textit{ans.})$$

PROBLEMS AND SOLUTIONS

5.03 Expressed on a "as fired" basis, a coal shows the following ultimate analysis:

C	S	H_2	O_2	moisture
75.8	1.5	4.8	8.9	0.5

What is the net heating value of this coal expressed in terms of kcal/kg?

$$H_\nu = (80.8)(75.8) + \left[\left(287\ 4.8 - \frac{8.9}{8}\right)\right] + [22.45(1.5)] - [6(0.5)]$$

$$= 7216.6 \text{ kcal/kg} \quad (ans.)$$

5.04 The specific heat consumption of a kiln is stated at 5.2 million Btu/ton of clinker based on the gross heating value of 12,750 Btu/lb of coal. What is the specific heat consumption of this kiln in terms of net Btu/ton when the coal contains 4.3 percent hydrogen?

$$H_{\nu\,(net)} = 12{,}750 - [9270(0.043)] = 12351 \text{ Btu/lb}$$

and

$$5.2\left(\frac{12351}{12750}\right) = 5.04 \text{ million net Btu/ton clinker} \quad (ans.)$$

5.10 Given the following analysis on a loss free basis:

		CaO	SiO_2	Al_2O_3	Fe_2O_3
Ash	:	0.4	49.3	28.6	20.9
Kiln feed	:	67.9	19.8	6.9	2.1
Clinker	:	66.5	20.5	7.3	2.5

What percent ash does the clinker contain?

$$x_1 = \frac{66.5 - 67.9}{0.4 - 67.9} 100 = 2.07$$

$$x_2 = \frac{20.5 - 19.8}{49.3 - 19.8} 100 = 2.37$$

$$x_3 = \frac{7.3 - 6.9}{28.6 - 6.9} 100 = 1.84$$

$$x_4 = \frac{2.5 - 2.1}{20.9 - 2.1} 100 = 2.13$$

$$\text{Percent ash in clinker} = \frac{2.07 + 2.37 + 1.84 + 2.13}{4} = 2.10 \text{ percent}$$

(*ans.*)

Chapter 6

pH: HYDROGEN–ION–CONCENTRATIONS

6.01 Definition of pH

The pH value of a chemical is indicated by the negative log of the hydrogen-ion-concentration (h_{c^+}).

h_{c^+}	1	10^{-1}	10^{-2}	.....	10^{-7}	.....	10^{-13}	10^{-14}
pH	0	1	2	.....	7	.....	13	14
← acid —					neutral	— basic (alkaline) →		
$[H^+]$ dominant						$[OH^-]$ dominant		

In a compound, if:

$[H^+]$	>	$[OH^-]$	:	compound is acidic in nature
$[OH^-]$	>	$[H^+]$	:	compound is basic (alkaline)
$[H^+]$	=	$[OH^-]$	:	compound is neutral

6.02 Calculation of pH

$$\text{pH} = x - \log y$$

$$[H^+] = y10^{-x}$$

$$[H^+] \cdot [OH^-] = \text{constant } (\approx 10^{-14})$$

Example

A solution of NaOH is 0.015 N. What is the pH?

$$[OH^-] = 0.015 \text{ gram equivalent per liter}$$

$$\log [OH^-] = -1.824$$

$$[OH^-] = 10^{-1.824}$$

but

$$[H^+] \cdot [OH^-] = 10^{-14}$$

thus,

$$[H^+] = \frac{10^{-14}}{10^{-1.824}} = 10^{-12.176}$$

$$\text{pH} = 12.176$$

6.03 Indicators

Reagent	*Indicator*	*Color change*
Acids	Red phenolphthalein	Color free
	Blue litmus paper	Red
	Yellow methyl orange	Red
Bases	Red litmus paper	Blue
	Colorless phenolphthalein	Red
	Red methyl orange	Yellow

PART II

BURNING

Chapter 7

TECHNICAL INVESTIGATION OF KILN PERFORMANCE

Introduction

The significant formulas for a study of the kiln performance and efficiency are given. An engineer should follow the sequence in which the formulas are presented herein as many calculations require the results obtained from earlier computations.

To simplify the engineer's task, all the formulas are presented in the form of work sheets that can be used to arrange the study in an orderly fashion. At the conclusion, a summary sheet is also given to compile all the significant results of this study.

Data, formulas, and results can be presented either in English or metric system units by using the appropriate worksheets in this chapter.

These work sheets can also be used to perform studies of parts of the kiln system (e.g., the cooler operation). The reader should have no difficulties in selecting the appropriate formulas from the worksheets in these instances.

For a complete study, it is essential that the kiln data be selected during a time when the kiln operates in a stable and uniform manner.

* * *

7.01 Technical Information on Kiln Equipment

Plant Location: ______________________ Kiln: ____________

Kiln

Process: ______________________

Manufactured by: ______________________

Year placed in operation: ______________________

Types of clinker produced: ______________________

Types of fuel burned: ______________________

Primary air source: ______________________

Feeder type: ______________________ ____________

Type of dust collector: ______________________

Dust processing: ______________________

Preheater

Type: ______________________

Manufactured by: ______________________

Year: ______________________

Cooler

Type: ______________________

Manufactured by: ______________________

Year: ______________________

Fans

		Type fan	*Rated capacity*	rpm	Hp
Cooler fan	1				
	2				
	3				
	4				
	5				
	6				

	Type fan	*Rated capacity*	rpm	Hp
Cooler Exhaust				
Induced draft				
Recycle fan				
Nose ring				
Shell cooling				

Drives

	Type	Hp	*Motor* rpm
Kiln drive			
Auxiliary drive			
Cooler drive			

Other kiln equipment

Function	Type	Hp	
....	....	....	....
....	....	....	....
....	....	....	....
....	....	....	....

Date of investigation:

Tested by

Chapter 8

KILN PERFORMANCE AND EFFICIENCY (ENGLISH SYSTEM OF UNITS)

DATA NEEDED

Fuel Analysis (Oil or Coal)

All values to be shown on the "as fired" basis

A_A	=	Percent ash	=	. . . . .
A_H	=	Percent hydrogen	=	. . . . .
A_C	=	Percent carbon	=	. . . . .
A_N	=	Percent nitrogen	=	. . . . .
A_O	=	Percent oxygen	=	. . . . .
A_S	=	Percent sulfur	=	. . . . .
A_M	=	Percent moisture	=	. . . . .
A_q	=	Heat value (gross Btu/lb)	=	. . . . .

Coal Ash Analysis

B_{Si}	=	Percent SiO_2	=	. . . . .
B_{Al}	=	Percent Al_2O_3	=	. . . . .
B_{Fe}	=	Percent Fe_2O_3	=	. . . . .
B_{Ca}	=	Percent CaO	=	. . . . .

Kiln Feed Analysis

All oxides must be shown on a "loss free" basis

C_{Si}	=	Percent SiO_2	=	.....
C_{Al}	=	Percent Al_2O_3	=	.....
C_{Fe}	=	Percent Fe_2O_3	=	.....
C_{Ca}	=	Percent CaO	=	.....
C_{Mg}	=	Percent MgO	=	.....
C_{Na}	=	Percent NaO_2	=	.....
C_K	=	Percent K_2O	=	.....
C_S	=	Percent SO_3	=	.....
C_{Ig}	=	Percent loss on ignition	=	.....
C_M	=	Percent moisture	=	.....
C_{50}	=	Percent retained on No. 50 sieve	=	.....
C_{200}	=	Percent retained on No. 200 sieve	=	.....
C_C	=	Percent organic matter	=	.....
C_{sp}	=	Specific gravity (dry solids)	=	.....

Kiln Exit Gas Analysis

Determined by Orsat analysis

D_{CO_2}	=	Percent CO_2 (by weight)	=	.....
D_{CO}	=	Percent CO (by weight)	=	.....
D_O	=	Percent O_2 (by weight)	=	.....
D_N	=	Percent N_2 (by weight)	=	.....

Precipitator Outlet Gas Analysis

E_{CO_2}	=	Percent CO_2 (by weight)	=	.....
E_{O_2}	=	Percent O_2 (by weight)	=	.....

Ambient Air Data

F_T = Ambient air temperature (°F) = ...
F_H = Relative humidity (percent) = ...
F_{El} = Elevation (feet above sea level) = ...
F_{Bar} = Barometric pressure (in. Hg) = ...

Weights of Materials

W_{Cl} = Kiln output (tph) = ...
W_{dF} = Lb dry feed per ton of clinker (show actual feed rate) = ...
W_A = Lb fuel per ton of clinker (as fired) = ...
W_{Ch} = Total tons of chains = ...

Kiln Dust Analysis

All oxides must be shown on a "loss free" basis

G_{Si} = Percent SiO_2 = ...
G_{Al} = Percent Al_2O_3 = ...
G_{Fe} = Percent Fe_2O_3 = ...
G_{Ca} = Percent CaO = ...
G_K = Percent K_2O = ...
G_S = Percent SO_3 = ...
G_{Ig} = Percent loss on ignition = ...
G = Percent of dust that is returned to kiln (decimal) = ...

Kiln Dimensions

L_1	= Kiln length (ft)	=	.....
L_2	= Kiln diameter (ft)	=	.....
L_3	= Primary air nozzle diameter (ft)	=	.....
L_4	= Length of chain section (ft)	=	.....
L_5	= Total chain surface area (ft^2)	=	.....
L_6	= Effective kiln feed end throat diameter (ft)	=	.....
L_7	= Refractory lining thickness (in.)	=	.....
L_8	= Total feet of chains	=	.....
L_9	= Kiln shell thickness (in.)	=	.....
L_{10}	= Kiln slope (degrees)	=	.....

Temperatures

T_c	= Feed entering kiln (°F)	=	.....
T_{Sa}	= Secondary Air (°F)	=	.....
T_{Pa}	= Primary Air (°F)	=	.....
T_{Be}	= Feed end (°F)	=	.....
T_{St}	= Cooler stack (°F)	=	.....
T_{Cl}	= Clinker at cooler exit (°F)	=	......
T_F	= Fuel as fired (°F)	=	.....
T_{Z_1}	= Average shell temperature, lower third of kiln (°F)	=	.....
T_{Z_2}	= Average shell temperature, middle third of kiln (°F)	=	.....
T_{Z_3}	= Average shell temperature, upper third of kiln (°F)	=	.....
T	= Average kiln room temperature (°F)	=	.....

Volumes

V_{Pa}	= Primary air flow (SCFM @ 32°F, S.L.)	=	.....
V_{Ex}	= Excess cooler air (SCFM @ 32°F, S.L.)	=	.....
V_{Co}	= Total air into cooler (SCFM @ 32°F, S.L.)	=	.....

Other Parameters

P_h	= Hood draft (in. H_2O)	=	. . . . .
P_{Be}	= Feed end draft (in. H_2O)	=	. . . . .
P_{Ks}	= Kiln speed (RPH)	=	. . . . .
X_1	= Cooler length, grates only (ft)	=	. . . . .
X_2	= Cooler width, grates only (ft)	=	. . . . .
X_3	= Total area of openings in hood (ft^2)	=	. . . . .

Clinker Analysis

H_{Si}	= Percent SiO_2	=	. . . . .
H_{Al}	= Percent Al_2O_3	=	. . . . .
H_{Fe}	= Percent Fe_2O_3	=	. . . . .
H_{Ca}	= Percent CaO	=	. . . . .
H_{Mg}	= Percent MgO	=	. . . . .
H_S	= Percent SO_3	=	. . . . .
H_{Alk}	= Percent total alkalies (as NaO_2)	=	. . . . .
H_{Ig}	= Percent loss on ignition	=	. . . . .

CALCULATIONS

8.01 Amount of Feed Required to Produce One Ton of Clinker

The result obtained herein does not include any dust losses. The assumption also made is that all the coal ash enters the clinker and none leaves the kiln with the exit gas.

$a = 35.68 \ C_{Ca}$ = ...

$b = 41.80 \ C_{Mg}$ = ...

$c = 20.0 \ C_{Al}$ = ...

$d = 20.0 \ C_{Si}$ = ...

$e = 20.0 \ C_{Fe}$ = ...

$f = 1.508 \ C_{Si} + 7.06 \ C_{Al}$ = ...

subtotal : = ...

$g = \text{subtotal times } \dfrac{100 - H_{Ig}}{100}$ = ...

Subtract ash in clinker.

$h = W_a \left(\dfrac{A_A}{100}\right)\left(\dfrac{100}{100 - C_{Ig}}\right)$ – = ...

i = lb feed required/ton of clinker = ...

8.02 Dust Loss

Dust losses are expressed in terms of equivalent feed.

$k_1 = W_{dF} - i$ = ...

$k_2 = \dfrac{k_1}{W_{dF}}$ = ...

In the absence of reliable data for actual feed usage, W_{dF} can also be determined by the following formula which takes into account the difference in the loss on ignition between the kiln feed and the dust among the other factors.

$$W_{dF} = \frac{2000}{100 - C_{Ig}}\left[100 - H_{Ig} + W_D\,(100 - G_{Ig}) - \frac{h}{2000}\right] = \ .....$$

where

W_D = tons of dust wasted per ton of clinker

The factor "h" can be obtained from **8.01**, all other variables are given in the data sheet.

8.03 Potential Clinker Compounds and Clinker Factors

$$C_3S = 4.07\,H_{Ca} - (7.6\,H_{Si} + 6.72\,H_{Al} + 1.43\,H_{Fe} + 2.85\,H_S) = \ldots\ldots$$

$$C_2S = 2.87\,H_{Si} - 0.75\,C_3S = \ldots\ldots$$

$$C_3A = 2.65\,H_{Al} - 1.69\,H_{Fe} = \ldots\ldots$$

$$C_4AF = 3.04\,H_{Fe} = \ldots\ldots$$

$$LSF = \frac{100\,H_{Ca}}{2.8\,H_{Si} + 1.65\,H_{Al} + 0.35\,H_{Fe}} = \ldots\ldots$$

$$S/R = \frac{H_{Si}}{H_A + H_{Fe}} = \ldots\ldots$$

$$A/F = \frac{H_{Al}}{H_{Fe}} = \ldots\ldots$$

$$\text{Percent liquid} = 1.13\,C_3A + 1.35\,C_4AF + H_{Mg} + H_{Alk} = \ldots\ldots$$

For an accurate determination of the clinker compounds, the clinker analysis has to be adjusted as shown in Chapter 1, Problem **1.09 (b)**, prior to using the oxide values in the above calculations. This takes into account the amount of CaO that is combined with the sulfur as $CaSO_4$ and the free lime present in the clinker.

8.04 Theoretical Heat Required to Produce One Ton of Clinker

The result obtained represents only that amount of heat that must be imparted to the feed to turn it into one ton of clinker. It does not include heat losses in the system nor does it account for the heat that is recoverable from the hot clinker after the clinkering reaction is completed. The result, however, does take into account the exothermic reaction at clinkering temperature.

$$l = 14795\,H_{Al} + 23326.7\,H_{Mg} + 27524.07\,H_{Ca} - 18416.58\,H_{Si} - 2123.88\,H_{Fe} = \ldots\ldots \text{ Btu/ton}$$

8.05 Percent of Infiltrated Air at Kiln Feed End

Infiltration is expressed on the basis of the kiln exit gases being equal to 100 moles and the precipitator gases containing 100 + x moles, where x represents the moles of infiltrated air.

First, solve the following equation for x

$$x = \frac{E_{O_2}}{100}(100+x) + \frac{3.78\,E_{O_2}}{100}(100+x) - 478\,\frac{D_{O_2}}{100}$$

then, the percent infiltrated air is

$$n = \frac{100x}{100+x} = \ldots\ldots$$

8.06 Excess Air Present in the Kiln

$$m = \frac{189\,(2D_{O_2} - D_{CO})}{D_N - [1.89\,(2D_{O_2} - D_{CO})]} = \ldots\ldots$$

8.07 Combustion Air Required to Burn One Pound of Fuel

This formula applies only to liquid and solid fuels. To determine the combustion air required for gaseous fuels, use the formula given in **13.08**.

$$o = \left(1 + \frac{m}{100}\right) [0.11594\,A_C + 0.34783\left(A_H - \frac{A_O}{8}\right) + 0.04348\,A_S] = \ldots\ldots$$

8.08 Weight of Combustion Air per Minute Entering Kiln

$$w_1 = \frac{W_{Cl}}{60} W_A o = \ldots\ldots$$

(Note: for natural gas firing use formula given in **13.08**).

8.09 Air Infiltration at Hood

$$w_5 = (7.74)(0.75)\chi_3\,(0.081 P_h)^{1/2} = \ldots\ldots$$

$$Z = \text{Percent infiltration} = \left(\frac{w_5}{w_1}\right) = \ldots\ldots$$

8.10 Cooler Air Balance

This air balance is being established in terms of lb/min flow rates. Keep in mind too that flow volumes in the data sheet are given at a base temperature 32 F at zero elevation.

$$w_4 = 0.081\,V_{Pa} = \ldots\ldots$$

$$w_3 = w_1 - w_4 - w_5 = \ldots\ldots$$

$$w_2 = 0.081\,V_{Ex} = \ldots\ldots$$

$$w_t = 0.081\,V_{CO} = \ldots\ldots$$

$$w_y = w_t - w_2 - w_3 = \ldots\ldots$$

where w_y represents the amount of cooler air lost by leaks in the cooler, the amount of air removed from the cooler for the drying of the coal, and/or the amount of cooler air recuperated for the precalciner or raw grind.

Note

$$\frac{w_1}{w_3 + w_4 + w_5} = \approx 1.0 = \;.\,.\,.\,.\,.$$

The percent primary air is

$$\frac{w_4}{w_1}\,100 = \;.\,.\,.\,.\,.$$

The air utilization efficiency of the cooler can be expressed by

$$\text{efficiency} = \frac{w_1}{w_t}\,100 = \;.\,.\,.\,.\,.$$

Note: To obtain the true efficiency of coolers where part of the heated air is used for raw grinding, coal drying or as secondary air in the precalciner subtract this amount of air from w_y before using the above formula.

8.11 Products of Combustion (lb/ton)

CO_2 from fuel $= 0.0367\,A_C W_A$ $= \;.\,.\,.\,.\,.$

SO_2 from fuel $= 0.02\,A_S W_A$ $= \;.\,.\,.\,.\,.$

H_2O from fuel $= 0.09\,A_H\,W_A$ $= \;.\,.\,.\,.\,.$

N_2 from fuel $= \left[\frac{A_N}{100} + 3.3478\,(0.0267\,A_C + 0.01\,A_S + 0.08\,A_H - 0.01\,A_O)\right] W_A$ $= \;.\,.\,.\,.\,.$

Subtotal $= \;.\,.\,.\,.\,.$

Add excess air: $\frac{m}{100}$ (subtotal) $= \;.\,.\,.\,.\,.$

w_6 = total (lb/ton clinker) $= \;.\,.\,.\,.\,.$

Note: for natural gas firing use the formula given in 13.08.

Find "m" in **8.06**.

8.12 Weight of Gases from Slurry (lb/ton)

CO_2 from feed $= (1 + 0.5\,k_2)(0.44a + 0.5216b) = \ldots\ldots$

H_2O_{free} from feed $= \dfrac{100\,W_{dF}}{100 - C_M} - W_{dF} = \ldots\ldots$

H_2O_{chem} from feed $= (1 + k_2)f = \ldots\ldots$

W_7 = total $= \ldots\ldots$

Note: Assumption: The dust wasted has given off 50 percent of its CO_2 and all of its moisture (free and chemical combined).

Find "k_2" in **8.02** and "a," "b," "f" in **8.01**.

8.13 Total Weight of Kiln Exit Gases

The best method for determining the weight of kiln exit gases is by measuring the actual flow at the back end of the kiln and subtracting the percent of infiltrated air that enters through leaks at the feed end (see **8.05**). In many instances, such flow measurements are suspect in accuracy because it is often difficult to measure the build-up in the duct where the measurements are taken. The large size of the ducts in this area also require special probes and a multitude of test points to obtain a meaningful traverse in the duct. In cases where this testing method is suspect, it is advisable to calculate the weight of kiln exit gases by adding the products found in **8.11** and **8.12**.

CO_2 from fuel			
CO_2 from slurry	Total CO_2	=	.....
H_2O from fuel			
H_2O from slurry (incl. chem)	Total H_2O	=	.....
50 percent of SO_2 from fuel	Total SO_2	=	.....
N_2 from combustion	Total N_2	=	.....
Excess air (see **8.11**)	Total Excess Air	=	.....

w_{gm} = Total = ...

w_{gd} = w_{gm} – Total H_2O (above) = ...

Note: In the above calculations the assumption is made that 50 percent of the sulfur in the fuel leaves the kiln with the exit gas and the other half leaves with the clinker.

8.14 Percent Moisture in Kiln Exit Gases

$$\text{Percent moisture} = \left[\frac{w_{gm} - w_{gd}}{w_{gm}}\right] 100 = \;.....$$

The percent moisture in the exit gases can also be determined from the condensate collected when the flue gases were tested with a dry gas meter [see Chapter 30, Section **30.03 (b)**].

8.15 Density of Kiln Exit Gas

The values for CO_2, SO_2, etc., in the following formula are found in **8.13**. Since the formula given applies to gas at 32°F at sea level, the result has to be converted to prevailing temperatures and pressures by using the well-known, general gas law.

a) At 32 F *and atmospheric pressure.*

$$d_{o_1} = \frac{0.1234CO_2 + 0.1827SO_2 + 0.0503H_2O + 0.0781N_2 + 0.0807 \text{ excess air}}{w_{gm}}$$

$$d_{o_1} = \ldots\ldots \text{ lb/ft}^3$$

b) At prevailing pressures and temperature.

$$d_{o_2} = d_{o_1}\left[\frac{492}{T_{BE} + 460}\right]\left[\frac{14.7 + 0.0361P_{BE}}{14.7}\right] = \ldots\ldots$$

8.16 Volume of Moist Kiln Exit Gases

$$v_{Be} = \frac{w_{gm}}{d_{o_2}} = \ldots\ldots \text{ ft}^3\text{/ton clinker}$$

$$v_{ACFM} = \frac{W_{Cl}v_{Be}}{60} = \ldots\ldots \; ACFM$$

8.17 Kiln Performance Factors

These factors are often used to compare a given kiln performance with another of similar dimensions.

a) Cooler air factor.

Expresses the specific air flow rate per square foot of grate area.

$$\frac{V_{CO}}{X_1X_2} = \ldots\ldots \; SCFM \text{ air/ft}^2 \text{ of grate area}$$

b) Primary air velocity.

The velocity of the primary air at the burner tip has a profound influence on the flame length and shape. For direct coal fired kilns one must include the volume of water vapor present in the primary air to obtain the proper velocity at the burner tip. This water vapor volume can be calculated from the percent moisture in the coal and converting this into the actual cubic feet per minute by using the steam table given in Chapter 22, Section **22.05**. The resultant volume is then added to the flow rate calculated by the following formula:

Convert primary air flow, V_{PA}, from standard to actual conditions by using

$$ACFM = \frac{V_{PA}}{\left[\frac{F_{BAR}}{29.92}\right]\left[\frac{T_{PA} + 460}{530}\right]} = \ldots\ldots$$

$$\text{Primary air velocity} = \frac{ACFM}{\pi(\frac{1}{2} L_3)^2} = \ldots\ldots \text{ft/min}$$

c) Specific kiln surface area loading.

This expresses the kiln output in terms of daily tons per square foot of kiln surface area.

$$\text{Kiln surface area} = (L_2 - 0.1667\, L_7)\pi L_1 = \ldots\ldots \text{ft}^2$$

$$\text{Specific area loading} = \frac{24\, W_{Cl}}{\text{ft}^2 \text{ surface}} = \ldots\ldots \text{tpd/ft}^2$$

d) Specific kiln volume loading.

This factor expresses the kiln output in terms of daily tons per cubic foot of internal kiln volume. This factor is often used to determine if a given kiln operates at, below, or above the design capacity for this kiln.

$$\text{Kiln volume} = \pi(\tfrac{1}{2}L_2 - 0.0833 L_7)^2 L_1 = \ldots\ldots \text{ft}^3$$

$$\text{Specific volume loading} = \frac{24 W_{Cl}}{\text{ft}^3 \text{ volume}} = \ldots.. \text{ tpd/ft}^3$$

e) Specific thermal loading of the burning zone.

Experience with different types and sizes of kilns has shown that there exists a relationship between the specific thermal loading of the burning zone and the life expectancy of the refractory. The higher the specific thermal loading, the shorter the brick life.

$$\text{Burning zone surface} = \pi(L_2 - 0.1667 L_y)L_x = \ldots.. \text{ft}^2$$

$$\text{Specific thermal loading} = \frac{W_A W_{Cl} A_q}{\text{B.Z. surface}} = \ldots.. \text{ Btu/h/ft}^2$$

where

L_y = lining thickness (inches)

L_x = length of the burning zone

Burning zone length, for this purpose, is defined as that length over which a thick, dark coating is formed over the refractory.

f) Gas velocity in the calcining zone.

For wet kilns, $w_{gc} = w_{gm} - [0.9(H_2O_{\text{free}}$ from slurry)].

For all other kilns

$$w_{gc} = w_{gm}$$

$$v_{cal} = \frac{w_{gc} W_{Cl}}{72\,(0.5 L_2 - 0.0833 L_7)^2\,\pi} = \ldots.. \text{ ft/min}$$

8.18 Results of Kiln Performance Study

8.01	i	Theoretical amount of feed required	: .... lb/ton
	h	Ash content in clinker	: .... lb/ton
8.02	k_1	Amount of feed wasted as dust	: .... lb/ton
	k_2	Dust loss in terms of fresh feed	: .... t/ton
8.03		C_3S in clinker	: .... percent
		C_2S in clinker	: .... percent
		C_3A in clinker	: .... percent
		LSF, lime saturation factor	: ....
		S/R, silica ratio	: ....
		A/F, alumina-iron ratio	: ....
		Percent liquid	: .... percent
8.04	l	Theoretical heat required for clinker formation	: .... Btu/ton
8.05	n	Percent air infiltrated at feed end	: .... percent
8.06	m	Percent excess air in exit gas	: .... percent
8.07	o	Combustion air required (lb air/lb fuel)	: ....
8.08	w_1	Weight of combustion air required (lb air/min)	: ....
8.09	w_5	Air infiltrated at kiln hood	: .... lb/min
	z	Percent of air infiltration at kiln hood	: .... percent

8.10	w_t	Total air flow into cooler	: .... lb/min
	w_y	Cooler air lost due to leaks or for drying	: .... lb/min
	w_2	Excess air vented at cooler stack	: .... lb/min
	w_3	Secondary air flow to kiln	: .... lb/min
	w_4	Primary air flow	: .... lb/min
		Percent primary air	: .... percent
8.11	w_6	Total combustion products from fuel	: .... lb/ton cl.
8.12	w_7	Gases produced from feed or slurry	: .... lb/ton cl.
8.13	w_{gm}	Total moist exit gas flow	: .... lb/ton cl.
	w_{gd}	Total dry exit gas flow	: .... lb/ton cl.
8.14		Moisture content of exit gas	: .... percent
8.15	d_{o_1}	Density of moist exit gases at standard condition (32°F)	: .... lb/ft^3
	d_{o_2}	Density of moist exit gases at actual conditions	: .... lb/ft^3
8.16	v_{Be}	Volume of exit gas flow	: .... ft^3/ton cl.
	v_{ACFM}	Volume of exit gas flow in unit time	: .... ft^3/min
8.17		Cooler air factor (ft^3/ft^2 of grate area)	: .... ft^3/ft^2
		Primary air velocity at burner tip	: .... ft/min
		Specific surface loading of kiln	: .... tpd/ft^2
		Specific volume loading of kiln	: .... tpd/ft^3
		Thermal loading of burning zone	: .... $Btu/h/ft^2$

Chapter 9

KILN PERFORMANCE AND EFFICIENCY (METRIC SYSTEM OF UNITS)

DATA NEEDED

Fuel Analysis (Oil or Coal–as Fired)

A_A	= Percent ash	=	.....
A_H	= Percent hydrogen	=	.....
A_C	= Percent carbon	=	.....
A_N	= Percent nitrogen	=	.....
A_O	= Percent oxygen	=	.....
A_S	= Percent sulfur	=	.....
A_M	= Percent moisture	=	.....
A_Q	= Heat value (kcal/kg)	=	.....
A_J	= Heat value (kJ/kg)	=	.....

Coal Ash Analysis

B_{Si}	= Percent SiO_2	=	.....
B_{Al}	= Percent Al_2O_3	=	.....
B_{Fe}	= Percent Fe_2O_3	=	.....
B_{Ca}	= Percent CaO	=	.....

Kiln Feed Analysis (Loss Free Basis)

C_{Si}	= Percent SiO_2	=	. . . . .
C_{Al}	= Percent Al_2O_3	=	. . . . .
C_{Fe}	= Percent Fe_2O_3	=	. . . . .
C_{Ca}	= Percent CaO	=	. . . . .
C_{Mg}	= Percent MgO	=	. . . . .
C_{Na}	= Percent Na_2O	=	. . . . .
C_K	= Percent K_2O	=	. . . . .
C_S	= Percent SO_3	=	. . . . .
C_{Ig}	= Percent ignition loss	=	. . . . .
C_M	= Percent moisture	=	. . . . .
C_{4900}	= Percent –4900 mesh	=	. . . . .
C_{400}	= Percent +400 mesh	=	. . . . .
C_C	= Percent organics	=	. . . . .

Kiln Exit Gas Analysis (By Orsat)

D_{CO_2}	= Percent CO_2	(by weight)	=	. . . . .
D_{CO}	= Percent CO	(by weight)	=	. . . . .
D_O	= Percent O_2	(by weight)	=	. . . . .
D_N	= Percent N_2	(by weight)	=	. . . . .

Precipitator Outlet Gas Analysis

E_{CO_2}=	= Percent CO_2	(by weight)	=	. . . . .
E_{O_2}	= Percent O_2	(by weight)	=	. . . . .

Ambient Air and Location

F_T	= Ambient air temp. (°C)	=	. . . . .
F_H	= Percent relative humidity	=	. . . . .
F_{EL}	= Elevation (meters above sea level)	=	. . . . .
F_{Bar}	= Barometric pressure (mm Hg)	=	. . . . .

Weights of Materials

W_{Cl}	= Kiln output (kg/h)	=	. . . . .
W_{dF}	= Dry feed rate (kg/kg of clinker)	=	. . . . .
W_A	= Fuel rate, as fired (kg/kg of clinker)	=	. . . . .

Kiln Dust Analysis (Loss Free Basis)

G_{Si}	= Percent SiO_2	=	. . . . .
G_{Al}	= Percent Al_2O_3	=	. . . . .
G_{Fe}	= Percent Fe_2O_3	=	. . . . .
G_{Ca}	= Percent CaO	=	. . . . .
G_K	= Percent K_2O	=	. . . . .
G_S	= Percent SO_3	=	. . . . .
G_{Ig}	= Percent ignition loss	=	. . . . .
G	= Percent of collected dust that is returned to the kiln (decimal)	=	. . . . .

Kiln Dimensions

L_1	=	Kiln length (m)	=	. . . . .
L_2	=	Kiln diameter (m)	=	. . . . .
L_3	=	Effective burner tip orifice area (m^2)	=	. . . . .
L_4	=	Length of chain section (m)	=	. . . . .
L_5	=	Total chain surface area (m^2)	=	. . . . .
L_6	=	Effective kiln feedend throat area (m^2)	=	. . . . .
L_7	=	Refractory thickness (mm)	=	. . . . .
L_8	=	Sum total of (all) chain length (m)	=	. . . . .
L_9	=	Kiln shell thickness (mm)	=	. . . . .
L_{10}	=	Kiln slope (degrees)	=	. . . . .

Operating Parameters

P_h	=	Hood draft (mm H_2O)	=	. . . . .
P_{Be}	=	Feed end draft (mm H_2O)	=	. . . . .
P_{Ks}	=	Kiln speed (rpm)	=	. . . . .

Clinker Analysis (Loss Free Basis)

H_{Si}	=	Percent SiO_2	=	. . . . .
H_{Al}	=	Percent Al_2O_3	=	. . . . .
H_{Fe}	=	Percent Fe_2O_3	=	. . . . .
H_{Ca}	=	Percent CaO	=	. . . . .
H_{Mg}	=	Percent MgO	=	. . . . .
H_S	=	Percent SO_3	=	. . . . .
H_{Alk}	=	Percent alkalies (total as Na_2O)	=	. . . . .
H_{Ig}	=	Percent ignition loss	=	. . . . .

Temperatures (Celsius)

T_C	= Feed entering kiln	=	. . . . .
T_{Sa}	= Secondary air	=	. . . . .
T_{Pa}	= Primary air	=	. . . . .
T_{Be}	= Kiln exit gas	=	. . . . .
T_{St}	= Cooler stack	=	. . . . .
T_{Cl}	= Clinker at cooler exit	=	. . . . .
T_F	= Fuel as fired	=	. . . . .
T_{Z_1}	= Average shell, lower third	=	. . . . .
T_{Z_2}	= Average shell, middle third	=	. . . . .
T_{Z_3}	= Average shell, upper third	=	. . . . .
T	= Kiln room	=	. . . . .

Air Volumes (*Standard* m^3/s @ 0 C, 760 mm Hg)

V_{Pa}	= Primary air flow	=	. . . . .
V_{Ex}	= Cooler vent stack	=	. . . . .
V_{CO}	= Total air into cooler	=	. . . . .
V_{Be}	= Kiln exit	=	. . . . .

Dimensions

X_1	= Cooler length, grates, (m)	=	. . . . .
X_2	= Cooler width, grates, (m)	=	. . . . .
X_3	= Total effective area of hood opening (m^2) where ambient air is infiltrated	=	. . . . .

CALCULATIONS

For detailed descriptions and notes pertaining to the formulas given, refer to the previously given formulas in the English system of units.

9.01 Amount of Feed Required to Produce One Kilogram of Clinker

$a = 0.01784 \; C_{Ca}$ = ...

$b = 0.0209 \; C_{Mg}$ = ...

$c = 0.01 \; C_{Al}$ = ...

$d = 0.01 \; C_{Si}$ = ...

$e = 0.01 \; C_{Fe}$ = ...

$f = 0.00075 \; C_{Si} + 0.0035 \; C_{Al}$ = ...

Subtotal = ...

$$g = \text{subtotal}\left(\frac{100 - H_{Ig}}{100}\right)$$ = ...

Subtract ash in clinker

$$h = W_a \left(\frac{A_A}{100}\right)\left(\frac{100}{100 - C_{Ig}}\right)$$ – = ...

i = kg feed required per kg clinker = ...

9.02 Dust Loss

Dust losses are expressed in terms of equivalent feed.

$k_1 = W_{dF} - i$ = ... kg/kg cl.

$$k_2 = \frac{k_1}{W_{dF}} \qquad = \text{.....percent}$$

In the absence of reliable data for actual feed usage, W_{dF} can also be determined as follows:

$$W_{dF} = \frac{100}{100 - C_{Ig}} \left[W_D \left(\frac{100 - G_{Ig}}{100} \right) + \left(\frac{100 - H_{Ig}}{100} \right) - h \right] = \ldots\ldots$$

where

W_D = kg of dust wasted per kg clinker

"h" can be found in **9.01**.

9.03 Potential Clinker Compounds and Clinker Factors

$$C_3S = 4.07\,H_{Ca} - (7.6\,H_{Si} + 6.72\,H_{Al} + 1.43\,H_{Fe} + 2.85\,H_S) = \ldots\ldots$$

$$C_2S = 2.87\,H_{Si} - 0.75\,C_3S = \ldots\ldots$$

$$C_3A = 2.65\,H_{Al} - 1.69\,H_{Fe} = \ldots\ldots$$

$$C_4AF = 3.04\,H_{Fe} = \ldots\ldots$$

$$LSF = \frac{100\,H_{Ca}}{2.8\,H_{Si} + 1.65\,H_{Al} + 0.35\,H_{Fe}} = \ldots\ldots$$

$$S/R = \frac{H_{Si}}{H_{Al} + H_{Fe}} = \ldots\ldots$$

$$A/F = \frac{H_{Al}}{H_{Fe}} = \ldots\ldots$$

Percent liquid = $1.13\,C_3A + 1.35\,C_4AF + H_{Mg} + H_{Alk}$ = ...

9.04 Theoretical Heat Required to Produce One Kilogram Clinker

$$1 = 4.11\,H_{Al} + 6.48\,H_{Mg} + 7.646\,H_{Ca} - 5.116\,H_{Si} - 0.59\,H_{Fe} = \ldots\ldots \text{kcal/kg}$$

9.05 Percent of Infiltrated Air at Kiln Feed End

First, solve the following equation for x:

$$x = \frac{E_{O_2}}{100}(100 + x) + \frac{3.78\,E_{O_2}}{100}(100 + x) - 478\,\frac{D_{O_2}}{100}$$

The percent of infiltrated air is found by the following formula:

$$n = \frac{100x}{100 + x} = \ldots\ldots$$

9.06 Excess Air Present in the Kiln

$$m = \frac{189\,(2.0\,D_{O_2} - D_{CO})}{D_N - [1.89\,(2.0 D_{O_2} - D_{CO})]} = \ldots\ldots$$

9.07 Combustion Air Required to Burn One Kilogram of Fuel (Solid or Liquid)

$$o = \left(1 + \frac{m}{100}\right)\left[0.11594\,A_C + 0.34783\left(A_H - \frac{A_O}{8}\right) + 0.04348\,A_S\right] = \ldots\ldots \text{kg air/kg fuel}$$

Note: For gaseous fuel see **13.01** in Chapter 13.

9.08 Weight of Combustion Air Required per Hour

$$w_1 = W_{Cl} W_{AO} = \ldots\ldots \text{kg/h}$$

Note: For gaseous fuel see **13.01** in Chapter 13.

9.09 Air Infiltration at Firing Hood

$w_5 = 11{,}720.3\chi_3\,(1.157P_h)^{1/2} \quad = \ldots\ldots$ kg/h

$z = \text{Percent infiltration} = \dfrac{w_5}{w_1}\,100 \quad = \ldots\ldots$ percent

9.10 Cooler Air Balance

$w_4 = 4654.44\,V_{Pa} \quad = \ldots\ldots$ kg/h

$w_3 = w_1 - w_4 - w_5 \quad = \ldots\ldots$ kg/h

$w_2 = 4654.44\,V_{Ex} \quad = \ldots\ldots$ kg/h

$w_t = 4654.44\,V_{CO} \quad = \ldots\ldots$ kg/h

$w_y = w_t - w_2 - w_3 \quad = \ldots\ldots$ kg/h

Note:

$$\frac{w_1}{w_3 + w_4 + w_5} = \approx 1.0 = \ldots\ldots$$

The percent primary air is:

$$\frac{w_4}{w_1}\,100 = \ldots\ldots$$

The air utilization efficiency of the cooler can be expressed by the following formula:

$$\text{efficiency} = \frac{w_1}{w_t} 100 = \ldots\ldots$$

9.11 Products of Combustion

CO_2 from fuel $= 0.03667\, A_C\, W_A$ $= \ldots\ldots$

SO_2 from fuel $= 0.02\, A_S W_A$ $= \ldots\ldots$

H_2O from fuel $= 0.09\, A_H W_A$ $= \ldots\ldots$

$$N_2 \text{ from fuel} = \left[\frac{A_N}{100} + 3.3478\,(0.0267 A_C + 0.01\, A_S + 0.08\, A_H - 0.01\, A_O)\right] W_A = \ldots\ldots$$

Subtotal: $= \ldots\ldots$

Add excess air: $\dfrac{m}{100}$ (subtotal) $= \ldots\ldots$

w_6 = Total $= \ldots\ldots$ kg/kg clinker

Note: For natural gas firing, use the formula **13.08** in Chapter 13.

9.12 Weight of Gases from the Feed

CO_2 from feed $= (1 + 0.5k_2)(0.44a + 0.5216b)$ $= \ldots\ldots$

$$H_2O_{free} \text{ from feed} = \frac{100 W_{dF}}{100 - C_m} - W_{dF} = \ldots\ldots$$

H_2O_{chem} from feed $= (1 + k_2)f$ $= \ldots\ldots$

w_7 = Total $= \ldots\ldots$ kg/kg cl.

Note: The assumption is made that the wasted dust has been 50 percent calcined. Find a, b, f in **9.01** and k_2 in **9.02**.

9.13 Total Weight of Kiln Exit Gases

Adding the products in **9.11** and **9.12** gives the total weight of exit gas.

CO_2 from fuel = ...
CO_2 from slurry = ...

Total CO_2 = ...

H_2O from fuel = ...
H_2O from feed = ...

Total H_2O = ...

50 percent of SO_2 from fuel : Total SO_2 = ...

N_2 from combustion : Total N_2 = ...

Excess air (see **9.11**) : Excess air = ...

w_{gm} = Total moist gases = ... kg/kg clinker

Dry gases = w_{gd} = w_{gm} – total H_2O = ... kg/kg

9.14 Percent Moisture in Kiln Exit Gas

$$\text{Percent } H_2O = \frac{w_{gm} - w_{gd}}{w_{gm}} 100 = \text{..... percent}$$

9.15 Density of Kiln Exit Gas

a) At 0 C, 760 mm Hg

$$d_{o_1} = \frac{1.977CO_2 + 2.927SO_2 + 0.806H_2O + 1.251N_2 + 1.2928 \text{ excess air}}{w_{gm}}$$

$$= \text{..... kg/Nm}^3$$

Note: Use calculated values from **9.13** for CO_2, SO_2, etc.

b) At prevailing pressures and temperatures

$$d_{o_2} = d_{o_1}\left(\frac{273.2}{T_{Be} + 273.2}\right)\left(\frac{760 - 0.0736P_{Be}}{760}\right) = \ldots\ldots \text{kg/m}^3$$

9.16 Volume of Moist Kiln Exit Gas

$$v_{Be} = \frac{w_{gm}}{d_{o_2}} = \ldots\ldots \text{m}^3/\text{kg cl.}$$

$$v_s = \frac{(v_{Be})(W_{Cl})}{3600} = \ldots\ldots \text{m}^3/\text{s}$$

9.17 Kiln Performance Factors

a) Cooler air factor.

$$\frac{V_{CO}}{X_1 X_2} = \ldots\ldots \text{m}^3\text{s/m}^2 \text{ grate area}$$

b) Primary air velocity.

$$\text{m}^2/\text{s}_{(\text{act})} = V_{PA}\ \frac{T_{PA} + 273.2}{273.2}\ \frac{760}{F_{Bar}} = \ldots\ldots$$

$$\text{Primary air velocity} = \frac{\text{m}^2/\text{s}_{(\text{act})}}{\pi(\frac{1}{2}L_3)^2} = \ldots\ldots \text{m/s}$$

c) Specific kiln surface area loading.

Kiln surface area $= (L_2 - 0.002\,L_7)\pi L_1 = \ldots\ldots \text{m}^2$

$$\text{Specific area loading} = \frac{24\,W_{Cl}}{1000\,(\text{m}^2\ \text{surface})} = \ldots\ldots \text{daily t/m}^2$$

d) Specific kiln volume loading.

Using the "inside lining" kiln volume, the specific volume loading in terms of daily metric tons production is calculated as follows

$$\text{Specific volume loading} = \frac{24\,W_{Cl}}{1000(\text{m}^3\ \text{volume})} = \ldots\ldots \text{daily t/m}^3$$

e) Specific thermal loading of the burning zone.

Burning Zone surface $= \pi(L_2 - 0.002L_y)L_x = \ldots\ldots \text{m}^2$

$$\text{Specific Thermal loading} = \frac{(W_A)(W_{Cl})(A_q)}{(\text{m}^2\ \text{surface})} = \ldots\ldots \text{kcal/h/m}^2$$

$$= \frac{(W_A)(W_{Cl})(A_j)}{(\text{m}^2\ \text{surface})} = \ldots\ldots \text{kJ/h/m}^2$$

where

L_y = lining and coating thickness (mm)
L_x = length of burning zone (m)

9.18 Results of Kiln Performance Study

9.01	i	Theoretical amount of feed required	. . . . .kg/kg cl.
	h	Ash content in clinker	. . . . .kg/kg cl.
9.02	k_1	Amount of feed wasted as dust	. . . . .kg/kg cl.
	k_2	Percent dust loss, in terms of fresh feed	. . . . .percent
	W_{dF}	Actual amount of feed used	. . . . .kg/kg cl.
9.03		C_3S in clinker	. . . . .percent
		C_2S in clinker	. . . . .percent
		C_3A in clinker	. . . . .percent
		C_4AF in clinker	. . . . .percent
		LSF, lime saturation factor	. . . . .percent
		S/R, silica ratio	. . . . .percent
		A/F, alumina–iron ratio	. . . . .percent
		Percent liquid	. . . . .percent

9.04	l	Theoretical heat required	. . . . .kcal/kg
9.05	n	Percent air infiltrated at feed end	. . . . .percent
9.06	m	Excess air present in exit gas	. . . . .percent
9.07	o	Combustion air required/kg fuel	. . . . .kg/kg
9.08	w_1	Combustion air required/unit time	. . . . .kg/h
9.09	w_5	Air infiltrated at hood	. . . . .kg/h
	z	Percent of combustion air infiltrated	. . . . .percent
9.10	w_t	Total air flow into cooler	. . . . .kg/h
	w_2	Excess air vented at cooler stack	. . . . .kg/h
	w_3	Combustion air into the kiln	. . . . .kg/h
	w_4	Primary air	. . . . .kg/h
		Percent primary air	. . . . .percent
9.11	w_6	Combustion products	. . . . .kg/kg cl.
9.12	w_7	Gases from slurry or feed	. . . . .kg/kg cl.
9.13	w_{gm}	Total moist exit gases	. . . . .kg/kg cl.
	w_{gd}	Total dry exit gases	. . . . .kg/kg cl.
9.14		Moisture content in exit gas	. . . . .percent

9.15	d_{o_1}	Density of moist gases (std. condition)	. . . . .kg/m^3
	d_{o_2}	Density of moist gases (actual condition)	. . . . .kg/m^3
9.16	v_{Be}	Volume of exit gases	. . . . .m^3/kg cl.
	v_s	Exit gas volume per unit time	. . . . .m^3/s
9.17		Cooler air factor	. . . .$m^3 \cdot s/m^2$ grate
		Primary air velocity	. . . . .m/s
		Specific surface loading	. . . . .daily tons/m^2
		Specific volume loading	. . . . .daily tons/m^3
		Thermal loading of burning zone	. . . . .kcal/h/m^2

Chapter 10

HEAT BALANCE
(ENGLISH SYSTEM OF UNITS)

In Chapter 12 there are graphs that show the mean specific heat for gases and solids at stated temperatures. In all the formulas given, "Q" denotes the heat content (Btu/ton clinker) and "c_m" denotes the mean specific heat of the gas or solid at the stated temperature.

HEAT INPUT

10.01 Heat Input from Combustion of Fuel

$$Q = W_A A_q = \ldots\ldots \text{Btu/ton}$$

10.02 Heat Input from Sensible Heat in Fuel

$$Q = W_A c_m (T_F - 32) = \ldots\ldots \text{Btu/ton}$$

10.03 Heat Input from Organic Substance in Kiln Feed

It is assumed that the organic matter in the kiln feed has a constant heat value of 9050 Btu/lb.

$$Q = g\left(\frac{C_c}{100}\right) 9050 = \ldots\ldots \text{ Btu/ton}$$

(find g in **8.01**).

10.04 Heat Input from Sensible Heat in Kiln Feed

$$Q_1 = W_{dF}\, c_m\, (T_c - 32) = \ldots\ldots$$

$$Q_2 = (H_2O \text{ in slurry})(T_C - 32) = \ldots\ldots$$

$$Q, \text{Total} = \ldots\ldots \text{Btu/ton}$$

10.05 Heat Input from Cooler Air Sensible Heat

$$Q = w_t\left(\frac{60}{W_{Cl}}\right)(c_m)\,(F_t - 32) = \ldots\ldots \text{Btu/ton}$$

(find w_t in **8.10**).

10.06 Heat Input from Primary Air (Sensible Heat)

Make sure to include in this calculation only that amount of primary air that originates from ambient source. Do not include the fraction of primary air that has its origin from the cooler air.

$$Q = w_4\left(\frac{60}{W_{Cl}}\right)c_m\,(T - 32) = \ldots\ldots \text{Btu/ton}$$

(find w_4 in **8.10**).

10.07 Heat Input from Infiltrated Air (Sensible Heat)

When the temperature in the area of the firing hood is significantly different from "*T*", use the appropriate temperature in the ensuing calculation.

$$Q = w_5\left(\frac{60}{W_{Cl}}\right)c_m\ (T - 32) = \ldots\ldots \text{Btu/ton}$$

(find w_5 in **8.09**).

Note: The fuel efficiency, i.e., the specific heat consumption of the kiln under investigation, is given by the result obtained in **10.01**. One must keep in mind that this value is expressed in terms of gross Btu per ton of clinker. A footnote should be used to state whether the fuel efficiency is expressed in net or gross Btu per ton clinker to avoid confusion when this fuel efficiency is compared with European kilns that state this efficiency exclusively as net Btu per ton.

HEAT OUTPUT

10.08 Heat Required for Clinker Formation

Obtain this directly from **8.04**.

$$Q = \ldots\ldots \text{Btu/ton}$$

10.09 Heat Loss with Exit Gas

The heat loss in the exit gas is calculated from the heat content of each individual gas component. The weight of each component can be obtained from **8.13**.

$$Q_{CO_2} = w_{CO_2} c_m (T_{BE} - 32) = \dots\dots$$
$$Q_{H_2O} = w_{H_2O} c_m (T_{BE} - 32) = \dots\dots$$
$$Q_{SO_2} = w_{SO_2} c_m (T_{BE} - 32) = \dots\dots$$
$$Q_{N_2} = w_{N_2} c_m (T_{BE} - 32) = \dots\dots$$
$$Q_{\text{excess air}} = w_{\text{excess air}} c_m (T_{BE} - 32) = \dots\dots$$

Total: $Q = \dots\dots$ Btu/ton

10.10 Heat Loss Due to Moisture in Feed or Slurry

In **8.13**, the total amount of water, including the chemically combined water, has been calculated. This weight is used in the following calculation. Keep in mind that the result obtained represents only the amount of heat that has to be expanded to turn the given weight of water into steam at 32 F. The heat losses associated with raising this steam to the kiln exit gas temperature have already been included in **10.09**.

$$Q = (w_{\text{total } H_2O})\ 1075.8 = \dots\dots \text{ Btu/ton}$$

10.11 Heat Losses Due to Dust in Exit Gases

$$Q = k_1 c_m (T_{BE} - 32) = \dots\dots \text{ Btu/ton}$$

(find k_1 in **8.02**).

10.12 Heat Loss with Clinker at Cooler Discharge

$$Q = 2000 c_m (T_{Cl} - 32) = \dots\dots \text{ Btu/ton}$$

10.13 Heat Loss at Cooler Stack

$$Q = w_2 \frac{60}{W_{Cl}} c_m (T_{ST} - 32) = \ldots\ldots \text{Btu/ton}$$

(find w_2 in **8.10**).

10.14 Radiation and Convection Losses on Kiln Shell

The shell temperature is measured over the entire length of the kiln in five-foot intervals by means of a radiation pyrometer. The total kiln length is then divided into three equal areas and the average shell temperature, T_Z, is calculated for each zone. In Chapter 13, a graph is given that shows the heat transfer coefficient, φ , to be used in the following calculations.

Let

s_s = kiln shell surface area,

$$s_s = \pi L_2 \left(\frac{L_1}{3}\right) = \ldots\ldots \text{ft}^2$$

$$Q_1 = s_s \varphi (T_{Z_1} - T) \frac{1}{W_{Cl}} = \ldots\ldots$$

$$Q_2 = s_s \varphi (T_{Z_2} - T) \frac{1}{W_{Cl}} = \ldots\ldots$$

$$Q_3 = s_s \varphi (T_{Z_3} - T) \frac{1}{W_{Cl}} = \ldots\ldots$$

Total: $Q = \ldots\ldots$ Btu/ton

It is advisable to plot the temperatures found on graph paper to have a record available that shows potential weak areas in the lining or any significant changes in temperatures when the kiln is tested again at a later time.

10.15 Heat Loss Due to Calcination of Wasted Dust

This heat loss refers only to that fraction of dust which is wasted and *not* returned to the kiln.

Calculate first the percent calcination of the kiln dust:

$$\% = \frac{C_{Ig} - G_{Ig}}{C_{Ig}} = \;.\,.\,.\,.\,.$$

Second, calculate the total carbonates contained in the wasted dust:

$$TC_{\text{dust}} = \frac{a+b}{i} W_{dF} k_2 = \;.\,.\,.\,.\,.\ \text{lb}$$

The heat loss due to calcination of the wasted dust is determined by

$$Q = \%\ TC_{\text{dust}}\,(685.1) = \;.\,.\,.\,.\,.\ \text{Btu/ton}$$

(find a, b, i in **8.01** and k_2 in **9.02**).

Heat Balance

Heat input	Btu/ton	Percent	Heat output	Btu/ton	Percent
10.01: Combustion of fuel			10.08: Theoretical heat required		
10.02: Sensible heat in fuel			10.09: Exit gas losses		
10.03: Organic matter in feed			10.10: Evaporation of moisture		
10.04: Sensible heat in feed			10.11: Dust in exit gas		
10.05: Sensible heat in cooler air			10.12: Clinker discharge		
10.06: Sensible heat in primary air			10.13: Cooler stack losses		
10.07: Sensible heat in infiltrated air			10.14: Kiln shell losses		
			10.15: Losses due to calcination of wasted dust		
			Unaccounted losses		
Total		100	Total		100

Note: Unaccounted losses are calculated by difference to make the two sides equal

Chapter 11

HEAT BALANCE
(METRIC SYSTEM OF UNITS)

In the appendix, the reader will find graphs for the mean specific heat of gases and solids that will be used in the ensuing calculations. In all the formulas given, "Q" denotes the heat content (kcal/kg), "Q_J" (kJ/kg), and "c_m" the mean specific heat in terms of (kcal/kg)(C), "c_j" in terms of (kJ/kg)(C).

HEAT INPUT

11.01 Heat Input from the Combustion of Fuel

$Q \quad = W_A A_q$ $\qquad$ = ... kcal/kg

$Q_J \quad = W_A A_j$ $\qquad$ = ... kJ/kg clinker

11.02 Heat Input from Sensible Heat in Fuel

$Q \quad = W_A c_m T_F$ $\qquad$ = ... kcal/kg

$Q_J \quad = W_A c_j T_F$ $\qquad$ = ... kJ/kg clinker

11.03 Organic Substance in Kiln Feed

It is assumed that the organic matter in the kiln feed has a constant heat value of 5028 kcal/kg and 21,036 kJ/kg.

$$Q = g \frac{C_c}{100} 5028 \qquad = \ldots\ldots \text{kcal/kg}$$

$$Q_J = g \frac{C_c}{100} 21{,}036 \qquad = \ldots\ldots \text{kJ/kg clinker}$$

(find g in **9.01**).

11.04 Heat Input from Sensible Heat in Kiln Feed

$$Q_1 = W_{dF} c_m T_C \qquad = \ldots\ldots$$

$$Q_2 = (\text{kg } H_2O \text{ in slurry}) \, T_C \qquad = \ldots\ldots$$

$$Q_{\text{total}} = \ldots\ldots \text{ kcal/kg}$$

$$Q_{J_1} = W_{dF} c_j T_C \qquad = \ldots\ldots$$

$$Q_{J_2} = (\text{kg} \cdot H_2O \text{ in slurry}) \, T_C \, 4.184 \qquad = \ldots\ldots$$

$$Q_{J\text{total}} = \ldots\ldots \text{ kJ/kg clinker}$$

11.05 Heat Input from Cooler Air Sensible Heat

$$Q = w_t \frac{1}{W_{Cl}} c_m F_T \qquad = \ldots\ldots \text{kcal/kg}$$

$$Q_J = w_t \frac{1}{W_{Cl}} c_j F_T \qquad = \ldots\ldots \text{kJ/kg clinker}$$

(find w_t in **9.10**).

11.06 Heat Input from Primary Air Sensible Heat

Include in this calculation only that amount of primary air which originates from the atmosphere. Do not include the fraction of primary air that has its origin from the cooler.

$$Q = w_4 \frac{1}{W_{Cl}} c_m F_T \quad = \ldots\ldots \text{kcal/kg}$$

$$Q_J = w_4 \frac{1}{W_{Cl}} c_j F_T \quad = \ldots\ldots \text{kJ/kg clinker}$$

(find w_4 in **9.10**).

11.07 Heat Input from Infiltrated Air Sensible Heat

When the temperature in the area where the majority of the infiltration takes place, is significantly different from "T", use the appropriate temperature for this calculation.

$$Q = w_5 \frac{1}{W_{Cl}} c_m T \quad = \ldots\ldots \text{kcal/kg}$$

$$Q_J = w_5 \frac{1}{W_{Cl}} c_j T \quad = \ldots\ldots \text{kJ/kg clinker}$$

(find w_5 in **9.09**).

HEAT OUTPUTS

11.08 Heat Required for Clinker Formation

For "Q" in terms of kcal/kg, the result of **9.04** can be entered here directly.

$$Q = \;.....\; \text{kcal/kg cl.}$$

In the International system of units (SI) this heat fraction is calculafed as follows:

$$Q_J = 61{,}902\, H_{Al} + 97{,}599\, H_{Mg} + 115{,}161\, H_{Ca} - 77{,}055\, H_{Si} - 8886\, H_{Fe} = \;.....\; \text{kJ/kg clinker}$$

11.09 Heat Loss with Kiln Exit Gas

The heat loss in the exit gas is calculated from the heat content of each individual gas component. The weights of these components has been calculated in **9.13**.

$$Q_{CO_2} = w_{CO_2} c_m T_{Be} = \;.....$$
$$Q_{H_2O} = w_{H_2O} c_m T_{Be} = \;.....$$
$$Q_{SO_2} = w_{SO_2} c_m T_{Be} = \;.....$$
$$Q_{N_2} = w_{N_2} c_m T_{Be} = \;.....$$
$$Q_{\text{excess air}} = w_{\text{excess air}}\, c_m\, T_{Be} = \;.....$$
$$\text{Total, } Q = \;.....\; \text{kcal/kg}$$

In terms of the International system of units (SI):

$$Q_{J(CO_2)} = w_{CO_2} c_j T_{Be} = \;.....$$
$$Q_{J(H_2O)} = w_{H_2O} c_j T_{Be} = \;.....$$
$$Q_{J(SO_2)} = w_{SO_2} c_j T_{Be} = \;.....$$
$$Q_{J(N_2)} = w_{N_2} c_j T_{Be} = \;.....$$
$$Q_{J(\text{excess air})} = w_{\text{excess air}}\, c_j\, T_{Be} = \;.....$$
$$\text{Total, } Q_J = \;.....\; \text{kJ/kg clinker}$$

11.10 Heat Loss Due to Moisture in Feed or Slurry

$Q = w_{\text{total } H_2O}\, 597.7$ = ... kcal/kg

$Q_J = w_{\text{total } H_2O}\, 2500.8$ = ... kJ/kg clinker

The total weight of water, ($w_{\text{total } H_2O}$) can be found in **9.13**. The results obtained represents only the amount of heat that has to be expanded to turn the given weight of water into steam at 0°Celsius. The heat losses associated with raising this steam to the kiln exit gas temperature have been included in **11.09**.

11.11 Heat Losses Due to Dust in the Kiln Exit Gases

$Q = k_1\, c_m\, T_{Be}$ = ... kcal/kg

$Q_J = k_1\, c_j\, T_{Be}$ = ... kJ/kg clinker

(find k_1 in **9.02**).

11.12 Heat Loss with Clinker at Cooler Discharge

$Q = c_m\, T_{Cl}$ = ... kcal/kg

$Q_J = c_j\, T_{Cl}$ = ... kJ/kg clinker

11.13 Heat Loss at Cooler Stack

$Q = \dfrac{w_2}{W_{Cl}}\, c_m\, T_{St}$ = ... kcal/kg

$Q_J = \dfrac{w_2}{W_{Cl}}\, c_j\, T_{St}$ = ... kJ/kg clinker

11.14 Heat Losses by Radiation on Kiln Shell

In the appendix find the heat transfer coefficients ϑ_m (kcal/m²h C) and ϑ_j (kJ/m²h C) for the average shell temperature, T_z, in each zone of the kiln.

Let

s_s = kiln shell surface area,

$$s_s = \pi L_2 \frac{L_1}{3} = \ldots\ldots \text{ m}^2$$

$$Q_1 = s_s \vartheta_m (T_{Z_1} - T) \frac{1}{W_{Cl}} = \ldots\ldots$$

$$Q_2 = s_s \vartheta_m (T_{Z_2} - T) \frac{1}{W_{Cl}} = \ldots\ldots$$

$$Q_3 = s_s \vartheta_m (T_{Z_3} - T) \frac{1}{W_{Cl}} = \ldots\ldots$$

Total, $Q = \ldots\ldots$ kcal/kg

In the International System of Units (SI)

$$Q_{J_1} = s_s \vartheta_j (T_{Z_1} - T) \frac{1}{W_{Cl}} = \ldots\ldots$$

$$Q_{J_2} = s_s \vartheta_j (T_{Z_2} - T) \frac{1}{W_{Cl}} = \ldots\ldots$$

$$Q_{J_3} = s_s \vartheta_j (T_{Z_3} - T) \frac{1}{W_{Cl}} = \ldots\ldots$$

Total, $Q_J = \ldots\ldots$ kJ/kg clinker

11.15 Heat Loss Due to Calcination of Wasted Kiln Dust

Calculate first the percent calcination of the kiln dust:

$$¢ = \frac{C_{Ig} - G_{Ig}}{C_{Ig}} = \ldots\ldots$$

Second, calculate the total carbonates in the kiln dust:

$$TC_{\text{dust}} = \frac{a + b}{i} W_{dF} k_2 = \ldots\ldots \text{kg}$$

Then,

$$Q = ¢\ TC_{\text{dust}} 380.6 = \ldots\ldots \text{kcal/kg}$$

$$Q_J = ¢\ TC_{\text{dust}} 1592.5 = \ldots\ldots \text{kJ/kg clinker}$$

Note: Include this neat loss in the heat balance only for that fraction of the dust that is wasted and *not* returned to the kiln.

(find a, b, i in **9.01** and k_2 in **9.02**).

HEAT BALANCE

Heat Input	kcal/kg	kJ/kg	Percent
11.01 Combustion of fuel			
11.02 Sensible heat, fuel			
11.03 Organics in feed			
11.04 Sensible heat, kiln feed			
11.05 Sensible heat, cooler air			
11.06 Sensible heat, primary air			
11.07 Sensible heat, infiltrated air			
Total:			100

Heat Output	kcal/kg	kJ/kg	Percent
11.08 Theoretical heat required			
11.09 Exit gas losses			
11.10 Evaporation			
11.11 Dust in exit gas			
11.12 Clinker discharge			
11.13 Cooler stack			
11.14 Kiln shell			
11.15 Calcination of waste dust			
Unaccounted losses			
Total:			100

Note: Unaccounted losses are calculated by difference.

Chapter 12

TECHNICAL INVESTIGATION OF THREE KILN MODELS

Introduction

Models of a wet process, a dry process, and a suspension preheater kiln are given here and their performance characteristics have been calculated in accordance to the formulas given in this chapter. The data are selected values of kiln parameters typical to these types of kilns when they are operated efficiently and properly maintained.

Data

Fuel analysis

(applied to all three kilns and stated on a "as fired" basis)

A_a = 7.7		A_o = 9.1	
A_h = 4.8		A_s = 2.1	
A_c = 74.4		A_m = 0.5	
A_n = 1.4			

A_q = 13,240 Btu gross/lb, 12790 Btu net/lb, 7106 kcal/kg

Kiln Feed Analysis (loss free)

	Wet kiln	Dry kiln	Preheater
C_{Si} =	21.1	20.5	21.3
C_{Al} =	6.5	6.7	6.5
C_{Fe} =	2.9	3.1	2.8
C_{Ca} =	66.4	66.9	66.8
C_{Mg} =	1.5	1.5	1.2
C_{Na} =	0.3	0.3	0.3
C_K =	0.85	0.55	0.45
C_S =	0.1	0.1	0.1
C_{Ig} =	35.3	35.5	35.6
C_M =	32.5	0.8	0.4
C_{sp} =	2.70	2.71	2.69

Kiln Exit Gas Analysis

	Wet kiln	Dry kiln	Preheater
D_{CO_2} =	26.4	26.3	28.1
D_{CO} =	–	–	–
D_O =	1.1	1.3	1.2
D_N =	72.5	72.4	70.7

Ambient Air (applies to all three kilns)

	English	Metric
F_T =	72	22
F_{Bar} =	29.65 in. Hg	753 mm Hg

Weights

		Wet kiln		*Dry kiln*		*Preheater*	
		English	Metric	English	Metric	English	Metric
W_{Cl}	=	31.5	28580	75.9	68860	121.5	110220
W_{dF}	=	3220	1.61	3140	1.57	3100	1.55
W_A	=	366	0.183	313	0.157	225	0.113

Kiln Dust

		Wet kiln	*Dry kiln*	*Preheater*
G_{Ig}	=	19.9	13.5	7.3
G	=	0.8	1.0	1.0

Kiln Dimensions

		Wet kiln		*Dry kiln*		*Preheater*	
		English	Metric	English	Metric	English	Metric
L_1	=	425	129.5	525	160.0	215	65.5
L_2	=	11.25	3.43	15.5	4.72	14.2	4.33
L_7	=	6	0.152	6	0.152	9	0.23

Temperatures

	Wet kiln		Dry kiln		Preheater	
	°F	°C	°F	°C	°F	°C
T_C =	75	24	125	52	105	41
T_{Pa} =	180	82	180	82	180	82
T_{Be} =	345	174	975	524	485	252
T_{St} =	285	141	485	252	515	268
T_{Cl} =	165	74	175	79	185	85
T_F =	75	24	75	24	75	24
T_{Z_1} =	565	296	585	307	560	293
T_{Z_2} =	455	235	485	252	510	265
T_{Z_3} =	225	107	395	201	175	79
T =	85	29	85	29	85	29

Volumes

	Wet kiln		Dry kiln		Preheater	
	English	Metric	English	Metric	English	Metric
V_{Pa} =	4195	1.87	9335	4.17	10640	4.75
V_{Co} =	45325	20.24	117795	52.60	178260	79.60
V_{Ex} =	17430	7.78	72106	32.18	122009	54.55

Other Parameters

		Wet kiln		Dry kiln		Preheater	
		English	Metric	English	Metric	English	Metric
P_h	=	−0.05	−1.27	−0.05	−1.27	−0.05	−1.27
X_1	=	50	15.2	94	28.7	135	41.1
X_2	=	7	2.13	8.5	2.59	9	2.74
X_3	=	114	0.074	260	0.168	225	0.145

Clinker Analysis

		Wet kiln	Dry kiln	Preheater
H_{Si}	=	21.8	20.83	21.7
H_{Al}	=	6.8	7.0	6.7
H_{Fe}	=	3.2	3.3	2.9
H_{Ca}	=	65.9	66.4	66.3
H_{Mg}	=	1.2	1.4	1.0
H_S	=	0.2	0.3	0.8
H_{Alk}	=	0.58	0.62	0.85
H_{Ig}	=	0.20	0.15	0.25

Summary of Kiln Performance Study Results

			Wet kiln	Dry kiln	Preheater
8.01	i	Theoretical amount of feed required	3072.0	3091.8	3088.9
	h	Ash content in clinker	43.6	37.4	26.9
8.02	k_1	Amount of feed wasted as dust	148.0	48.2	11.1
	k_2	Dust loss in terms of fresh feed	0.046	0.015	0.004
8.03		C_3S in clinker	51.7	59.4	53.5
		C_2S in clinker	23.6	15.0	21.9
		C_3A in clinker	12.6	13.0	12.9
		LSF, lime saturation factor	89.8	93.5	91.0
		S/R, Silica ratio	2.18	2.02	2.26
		A/F, Alumina–Iron ratio	2.13	2.12	2.31
		Percent liquid	29.11	29.76	28.31
8.04	l	Theoretical heat required for clinker formation	1,534,156	1,573,194	1,541,500
8.05	n	Percent air infiltrated at feed end	n.a.*	n.a.	n.a.
8.06	m	Percent excess air in exit gas	6.1	7.3	6.9

			Wet kiln	*Dry kiln*	*Preheater*
8.07	o	Combustion air required (lb air/lb fuel)	10.60	10.72	10.68
8.08	w_1	Weight of combustion air required (lb air/min)	2036.8	4244.5	4866.1
8.09	w_5	Air infiltrated at kiln hood	42.1	96.1	83.1
	z	Percent of air infiltration at kiln hood	2.1	2.3	1.7
8.10	w_t	Total air flow into cooler	3671.3	9541.4	14439.1
	w_y	Cooler air lost due to leaks or for drying	604.6	308.5	635.2
	w_2	Excess air vented at cooler stack	1411.8	5840.6	9882.7
	w_3	Secondary air flow to kiln	1654.9	3392.3	3921.2
	w_4	Primary air flow	339.8	756.1	861.8
		Percent primary air	16.7	17.8	17.7
8.11	w_6	Total combustion products from fuel	4239.7	3666.8	2625.6

			Wet kiln	*Dry kiln*	*Preheater*
8.12	w_7	Gases produced from feed or slurry	2731.7	1195.8	1167.8
8.13	w_{gm}	Total moist exit gas flow	6963.8	4856.0	3788.7
	w_{gd}	Total dry exit gas flow	5173.8	4616.1	3600.7
8.14		Moisture content of exit gas	25.7	4.9	5.0
8.15	d_{o_1}	Density of moist exit gases at std. condition (32°F)	0.0848	0.0938	0.0972
	d_{o_2}	Density of moist exit gases at actual conditions	0.0515	0.0319	0.0503
8.16	v_{Be}	Volume of exit gas flow	135,219	152,227	75,321
	v_{ACFM}	Volume of exit gas flow in unit time	70,990	192,567	152,525
8.17		Cooler air factor (ft^3/ft^2 of grate area)	129.5	147.4	146.7
		Primary air velocity at burner tip	n.a.	n.a.	n.a.
		Specific surface loading of kiln	0.055	0.076	0.34
		Specific volume loading of kiln	0.022	0.021	0.107
		Thermal loading of burning zone	80,800	98,900	99,400

*not available

HEAT BALANCE

(Wet Process Kiln)

	Heat input	1000's Btu/t	*Percent*		*Heat output*	1000's Btu/t	*Percent*
10.01:	Combustion of fuel	4845.84	96.5	10.08:	Theoretical heat required	1534.16	30.5
10.02:	Sensible heat in fuel	4.25	0.1	10.09:	Exit gas losses	646.45	12.9
10.03:	Organic matter in feed	–	–	10.10:	Evaporation of moisture	1925.64	38.3
10.04:	Sensible heat in feed	97.86	1.9	10.11:	Dust in exit gas	9.73	0.2
10.05:	Sensible heat in cooler air	65.17	1.3	10.12:	Clinker discharge	48.68	1.0
10.06:	Sensible heat in primary air	7.99	0.1	10.13:	Cooler stack losses	163.29	3.3
				10.14:	Kiln shell losses	582.71	11.6
10.07:	Sensible heat in infiltrated air	0.00	0.1	10.15:	Losses due to calcination of wasted dust	35.03	0.7
					Unaccounted losses	76.44	1.5
	Total	5022.11	100		Total	5022.11	100

Note: Unaccounted losses are calculated by difference to make the two sides equal.

HEAT BALANCE

(Dry Process Kiln)

	Heat input	Btu/t	*Percent*		*Heat output*	Btu/t	*Percent*
10.01:	Combustion of fuel	4144.12	96.5	10.08:	Theoretical heat required	1573.19	36.6
10.02:	Sensible heat in fuel	3.63	0.1	10.09:	Exit gas losses	1189.61	27.7
10.03:	Organic matter in feed	–	–	10.10:	Evaporation of moisture	258.10	6.0
10.04:	Sensible heat in feed	68.43	1.6	10.11:	Dust in exit gas	11.14	0.3
10:05:	Sensible heat in cooler air	70.30	1.6	10.12:	Clinker discharge	52.62	1.2
10.06:	Sensible heat in primary air	7.38	0.2	10.13:	Cooler stack losses	508.24	11.8
				10.14:	Kiln shell losses	521.17	12.1
10.07:	Sensible heat in infiltrated air	0.94		10.15:	Losses due to calcination of wasted dust	15.85	0.4
					Unaccounted losses	164.87	3.8
	Total	4294.80	100		Total	4294.80	100

Note: Unaccounted losses are calculated by difference to make the two sides equal.

HEAT BALANCE

(Suspension Preheater Kiln)

	Heat input	Btu/t	*Percent*		*Heat output*	Btu/t	*Percent*
10.01:	Combustion of fuel	2979.00	95.9	10.08:	Theoretical heat required	1541.5	49.6
10.02:	Sensible heat in fuel	2.61	0.1	10.09:	Exit gas losses	427.37	13.8
10.03:	Organic matter in feed	–	–	10.10:	Evaporation of moisture	202.21	6.5
10.04:	Sensible heat in feed	52.11	1.7	10.11:	Dust in exit gas	1.11	
10.05:	Sensible heat in cooler air	66.46	2.1	10.12:	Clinker discharge	56.61	1.8
10.06:	Sensible heat in primary air	5.26	0.2	10.13:	Cooler stack losses	527.80	18.4
				10.14:	Kiln shell losses	150.722	4.9
10.07:	Sensible heat in infiltrated air	0.51		10.15:	Losses due to calcination of wasted dust	5.321	0.2
					Unaccounted losses	148.302	4.8
	Total	3105.94	100		Total	3105.94	100

Note: Unaccounted losses are calculated by difference to make the two sides equal.

Chapter 13

SPECIFIC HEATS, HEAT TRANSFER COEFFICIENTS, AND COMPUTATIONS FOR NATURAL GAS FIRING

Compiled in this chapter, are the important parameters an engineer needs to complete a kiln investigation as outlined in Chapters 8 through 11.

Here the engineer will find the graphs that show him at a glance the appropriate specific heat and heat transfer coefficient to be used for his computations. The reader is advised to make use of the appropriate graphs and formulas in accordance with the particular system of units employed (English, metric, or S.I.) for this study. Failure to use the proper applicable system of units can introduce large errors in the result obtained.

The formulas shown in Chapter 8 through 9 apply to kilns fired with coal or fuel oil. In this chapter, the appropriate formulas for gas firing which should be used in Chapter 8 and 9 are also shown.

13.01 Mean Specific Heat of Clinker (Base: 0°C)

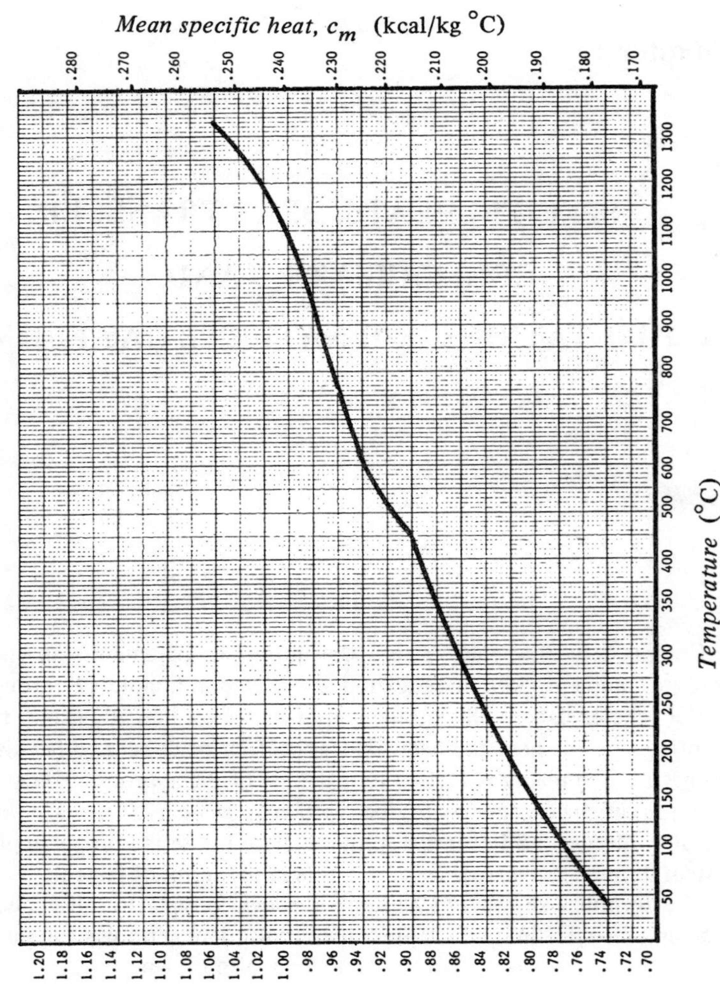

13.02 Mean Specific Heat of Raw Materials (Base: 0°C)

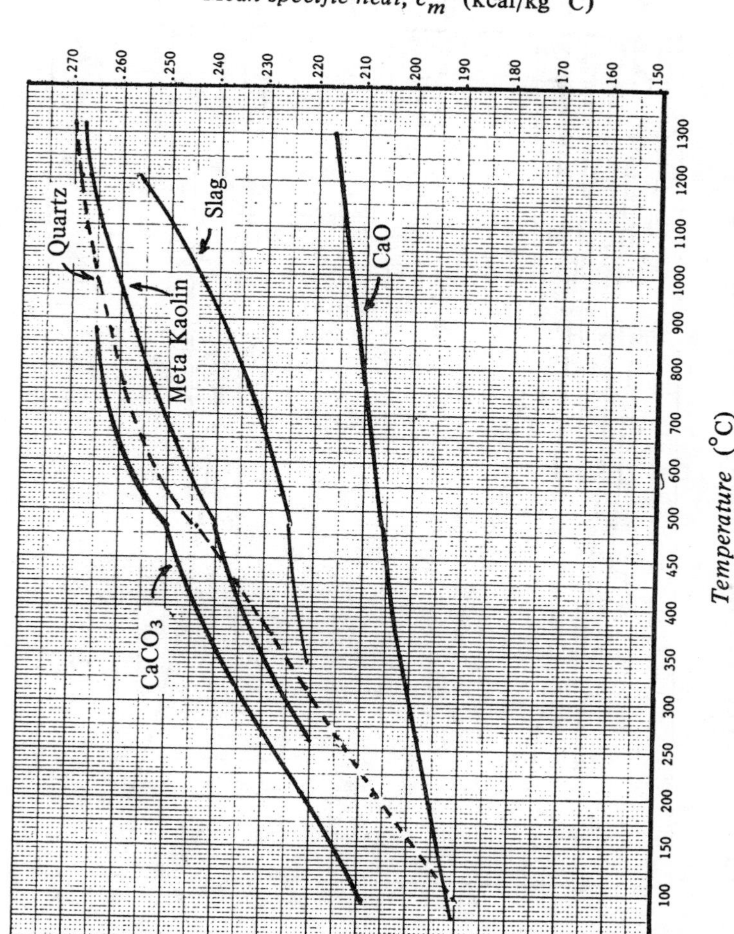

13.03 Mean Specific Heat of Exit Gas Components (Base: 0°C)

Mean specific heat, c_m (kcal/kg °C)

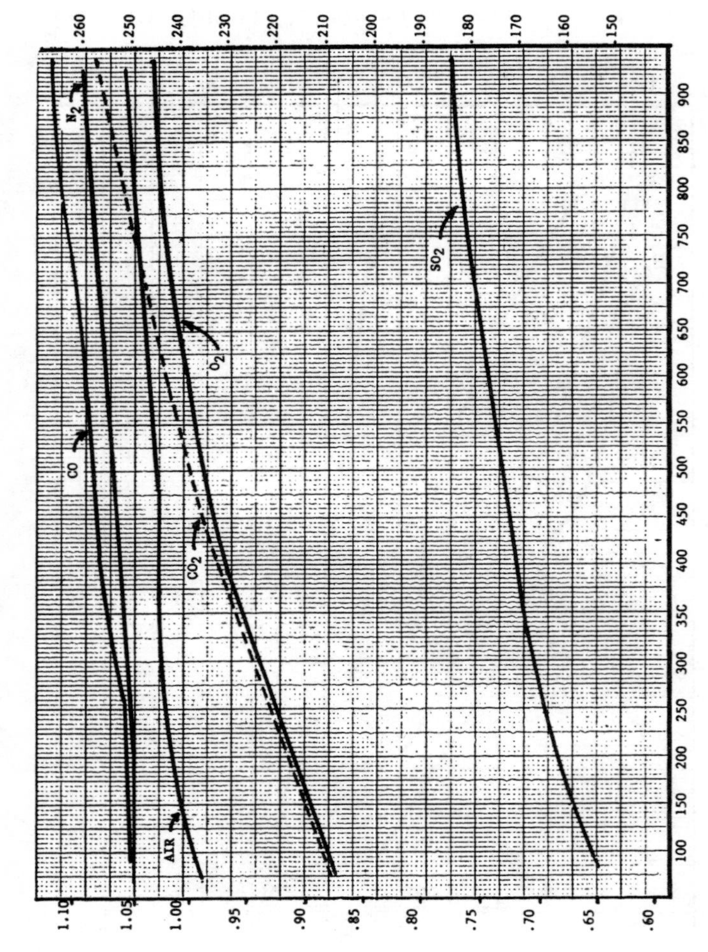

Temperature (°C)

Mean specific heat, c_i (kJ/kg °C)

13.04 Mean Specific Heat of Fuels (Base: 0°C)

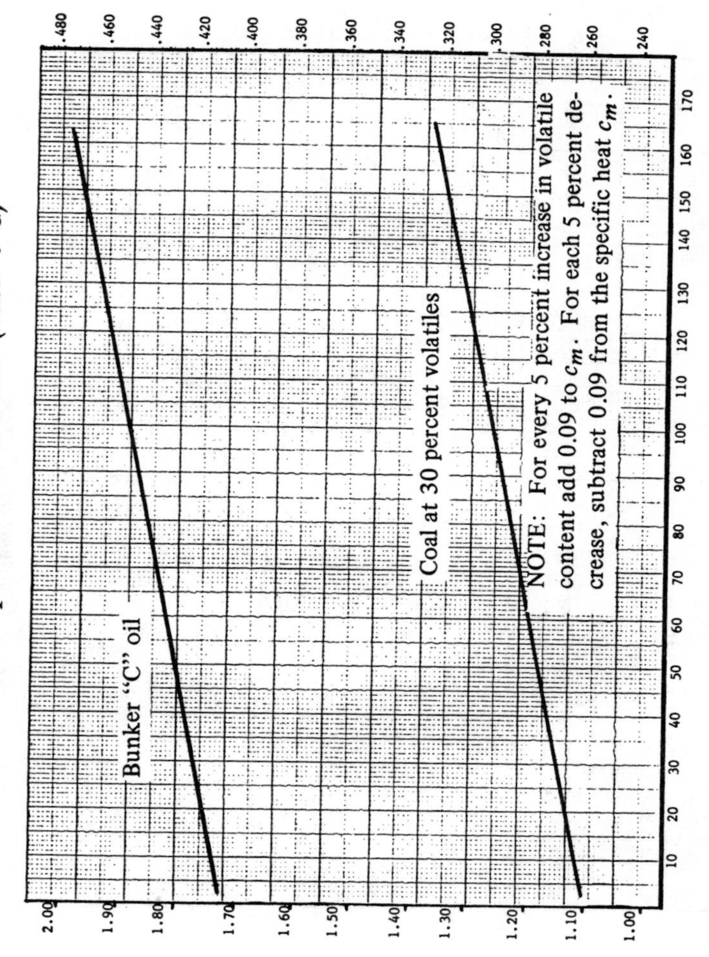

13.05 Mean Specific Heat of Water Vapor (Base: 0°C)

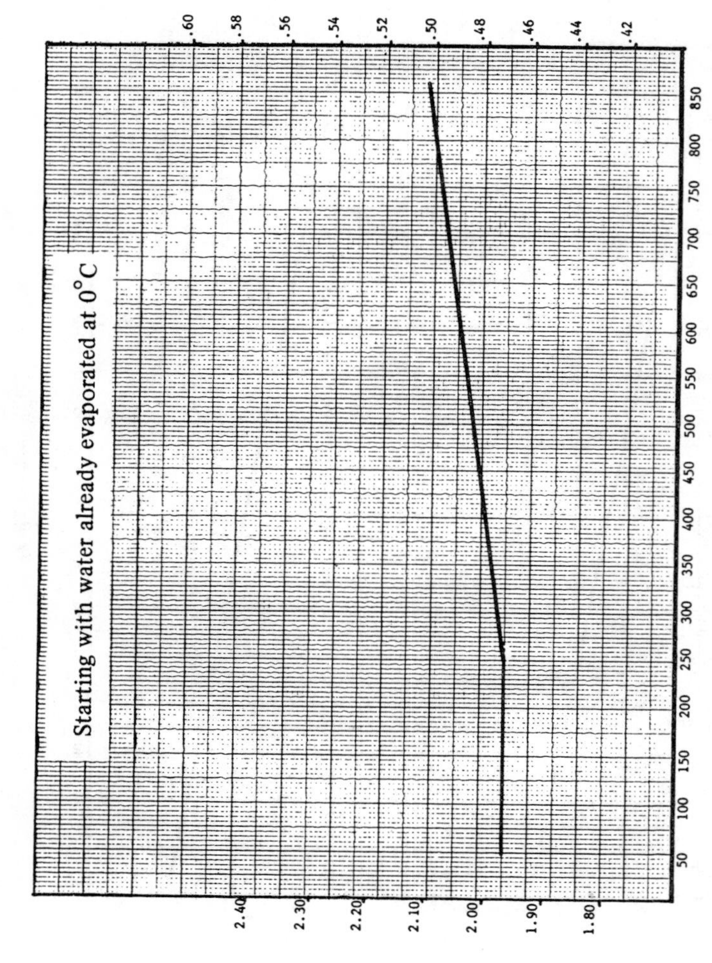

13.06 Heat Transfer Coefficients for Heat Loss on Kiln Shell

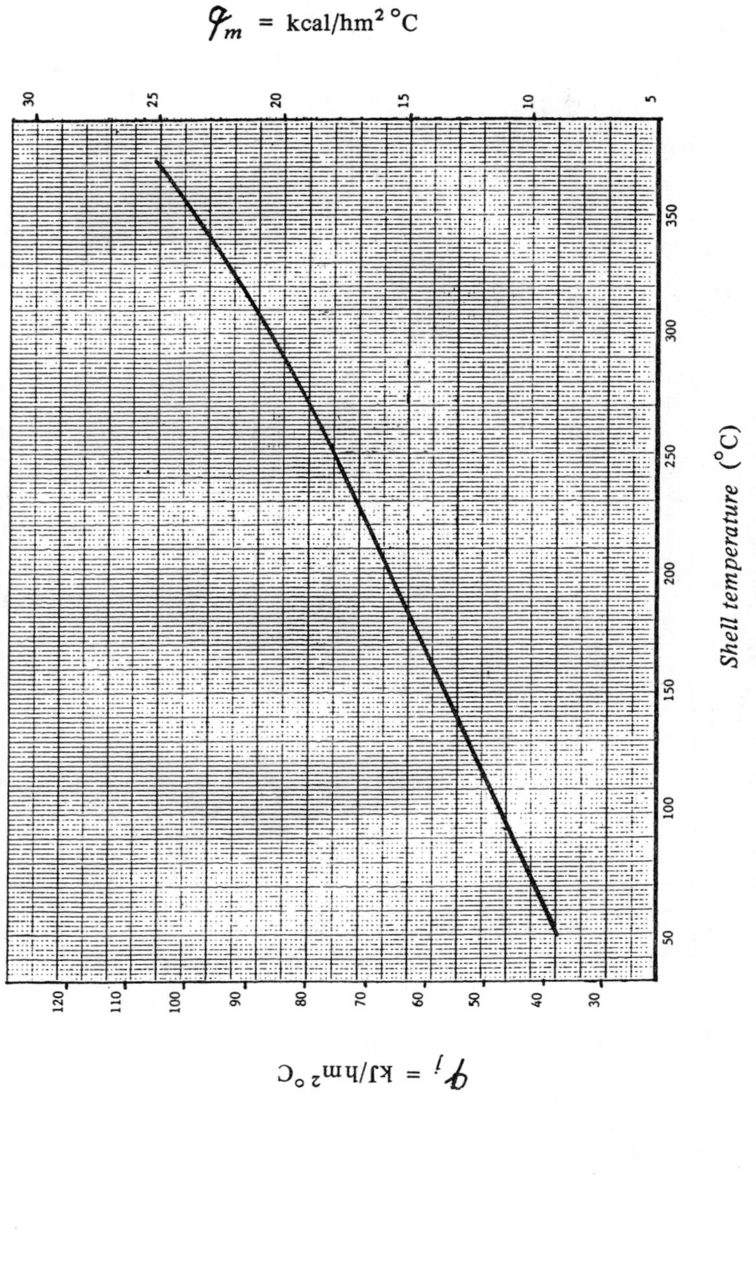

13.07 Mean Specific Heat of Clinker (Base: 32°F)

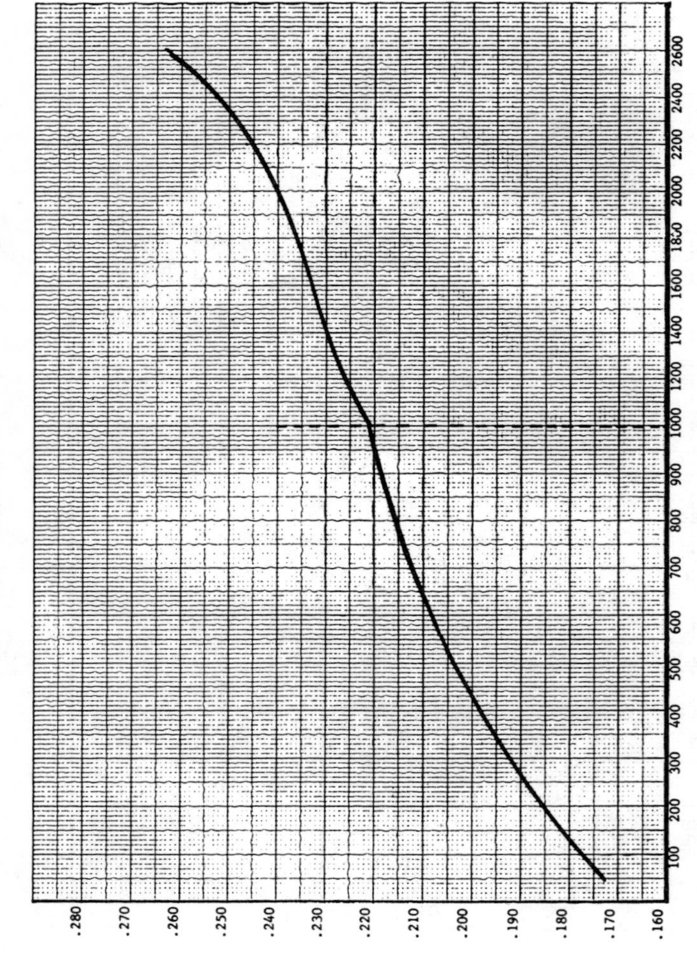

13.08 Mean Specific Heat of Cement Raw Materials (Base: 32°F)

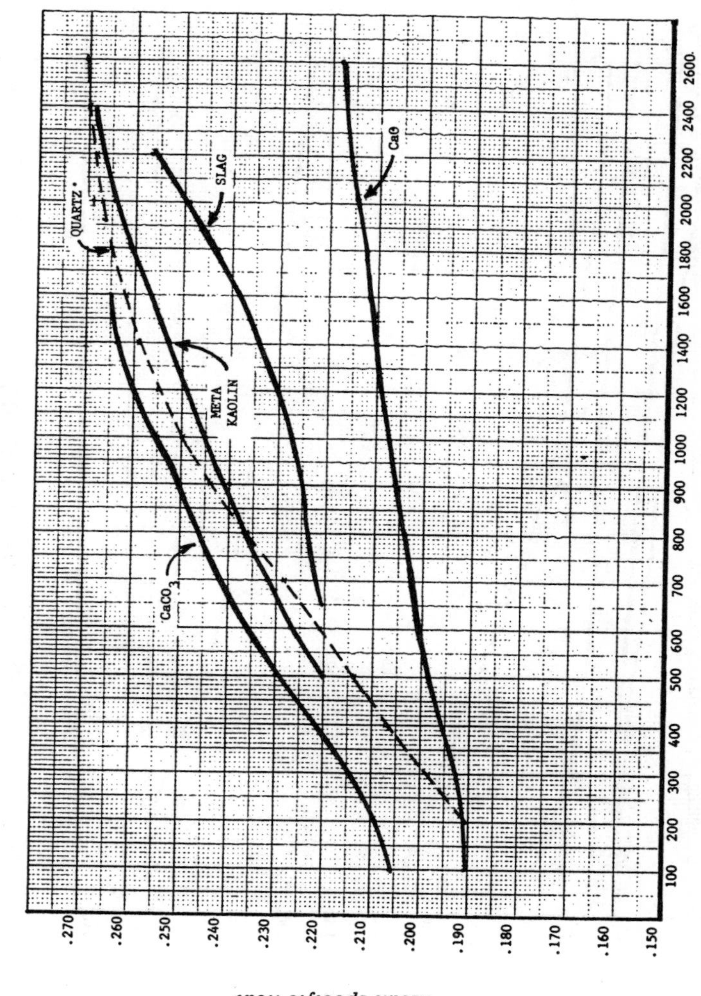

13.09 Mean Specific Heat of Exit Gas Components (Base: 32° F)

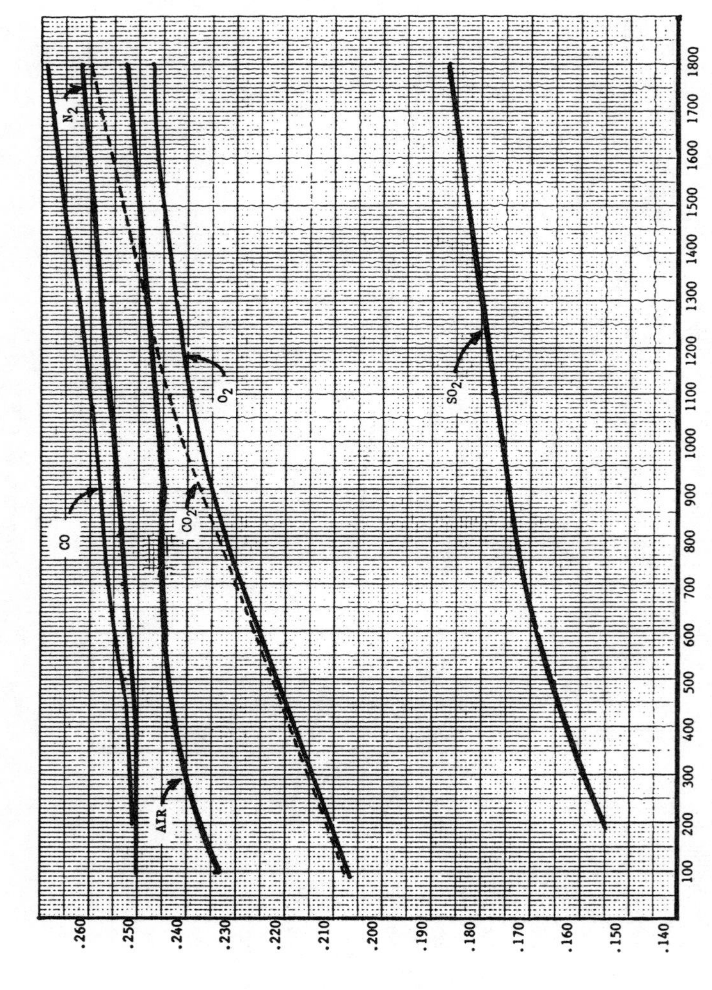

13.10 Mean Specific Heat of Fuels (Base: 32°F)

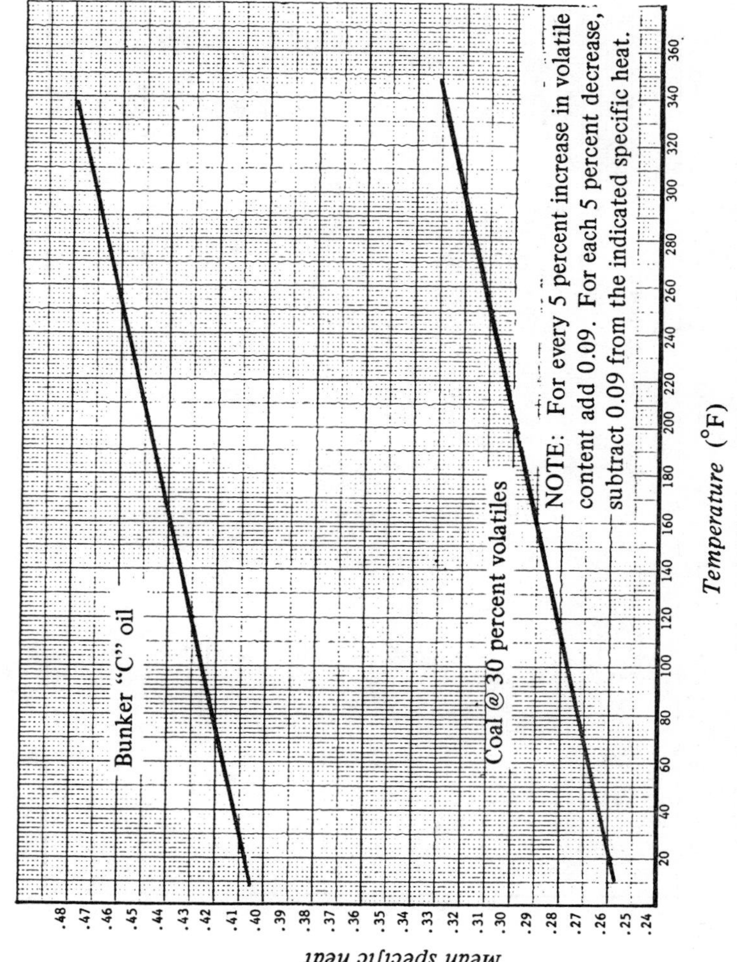

13.11 Mean Specific Heat of Water Vapor (Base: 32° F)

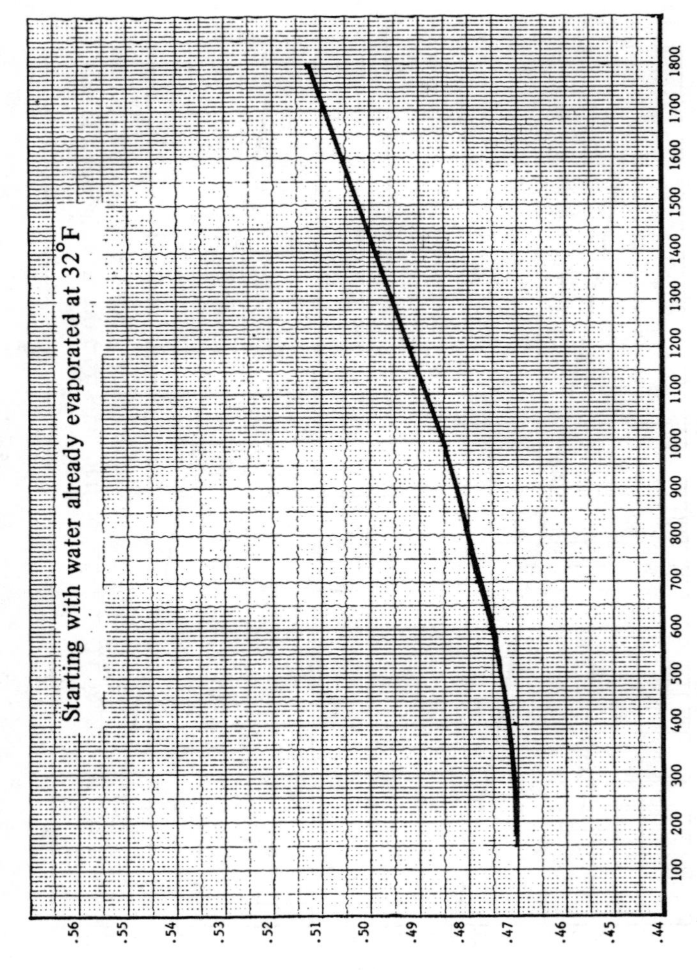

13.12 Heat Transfer Coefficient for Heat Losses on Kiln Shell, (q)

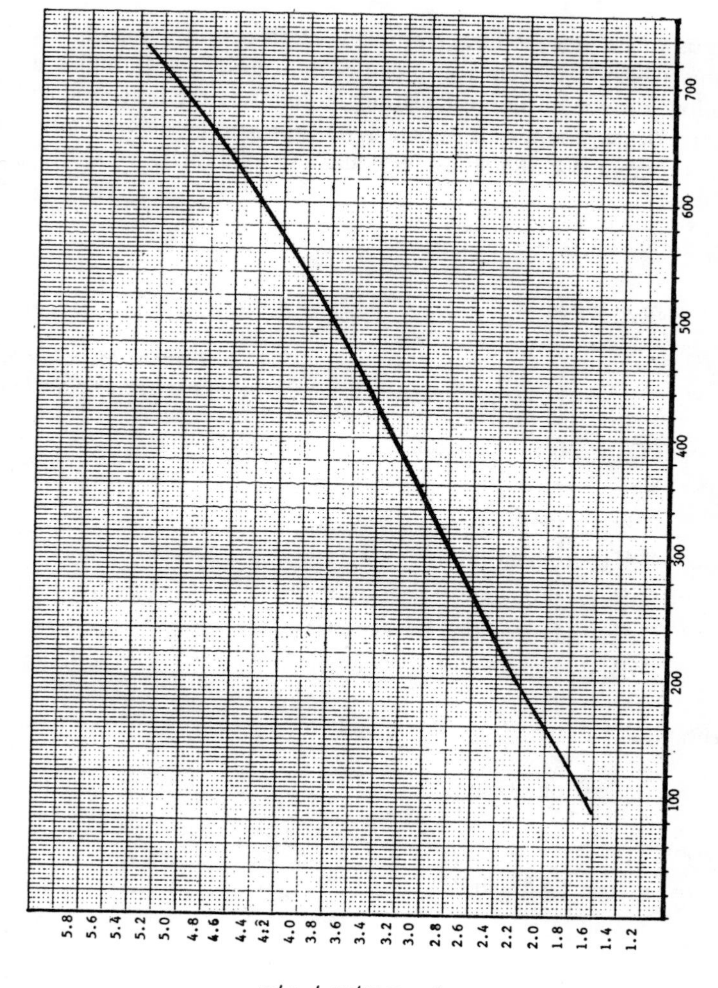

Shell temperature (°F)

q = Btu/ft²/°F/h

13.13 Computations for Natural Gas Firing

Analysis of natural gas fuels are usually expressed in terms of percent by volume which is the same as molar proportions. The formulas given below allow for combustion calculations in terms of the unit production of clinker. Hence, the results obtained are expressed in the same terms as the results computed in this study for liquid and solid fuels.

Data required:

Analysis of natural gas

			Percent by volume expressed as a decimal
a_1	= CO_2,	carbon dioxide	= .....
b_1	= N_2,	nitrogen	= .....
c_1	= CH_4,	methane	= ....
d_1	= C_2H_6,	ethane	= .....
e_1	= C_3H_8,	propane	= .....
f_1	= C_4H_{10},	butane (iso + *N*-butane)	= .....
g_1	= C_5H_{12+},	pentane (iso + *N*-pentane)	= .....

Fuel rate

W_a = ft^3 gas/ton of clinker = ...

or

= m^3 gas/kg of clinker = ...

Combustion air required (items **8.07** *and* **9.07** *for natural gas firing).*

When English units are used in this calculation, the result is also expressed in English units. Likewise, when metric units are used the results are expressed in metric terms.

$9.55c_1\ W_a$ = ...

$16.70d_1\ W_a$ = ...

$23.86e_1\ W_a$ = ...

$31.02f_1\ W_a$ = ...

$38.19g_1\ W_a$ = ...

Subtotal : = ...

o = Total air required = (subtotal) $\left(1 + \dfrac{m}{100}\right)$ = ... ft^3/ton

= ..... m^3/kg clinker

Weight of combustion air entering kiln (items **8.08** *and* **9.08** *for natural gas firing).*

English units:

$$w_1 = \left(\frac{W_{Cl}}{60}\right)0.081o = \ldots\ldots \text{ lb/min}$$

Metric units:

$$w_1 = (W_{Cl})1.2976o = \ldots\ldots \text{ kg/h}$$

Products of combustion (items **8.11** *and* **9.11** *for natural gas firing).*

a) English system of units

$$CO_2 \text{ from fuel} = W_a\,[0.123c_1 + 0.246d_1 + 0.368e_1 + 0.52f_1 + 0.602g_1 + 0.123a_1] = \ldots\ldots$$

$$H_2O \text{ from fuel} = W_a\,[0.1c_1 + 0.15d_1 + 0.196e_1 + 0.252f_1 + 0.315g_1 = \ldots\ldots$$

$$N_2 \text{ from fuel} = W_a\,[0.597c_1 + 1.048d_1 + 1.479e_1 + 1.92f_1 + 2.406g_1 + 0.078b_1] = \ldots\ldots$$

Subtotal = ...

Add excess air: $\dfrac{m}{100}$(subtotal) = ...

w_6 = total combustion product = ... lb/ton cl.

b) Metric system of units

$$CO_2 \text{ from fuel} = W_a\,[1.97c_1 + 3.94d_1 + 5.9e_1 + 8.33f_1 + 9.64g_1 + 1.97a_1] = \ldots\ldots$$

$$H_2O \text{ from fuel} = W_a\,[1.6c_1 + 2.4d_1 + 3.14e_1 + 4.04f_1 + 5.05g_1] = \ldots\ldots$$

$$N_2 \text{ from fuel} = W_a\,[9.55c_1 + 16.70d_1 + 23.86e_1 + 31.02f_1 + 38.19g_1 + 1.25b_1] = \ldots\ldots$$

Subtotal = ...

Add excess air: $\dfrac{m}{100}$ (subtotal) = ...

w_6 = total combustion product = ... kg/kg clinker

Chapter 14

USEFUL FORMULAS IN KILN DESIGN AND OPERATION

14.01 Cooling of Kiln Exit Gases by Water

Any moisture introduced into the gas stream is ultimately transferred into superheated steam and, in doing so, absorbs heat and cools the exit gases. The equations can be solved for any one of the unknowns if the other variables are known.

a) English system of units

$$w_2 0.248(t_1 - t_2) = w_1(1182.3 - t_3) + w_1(0.48)(t_2 - 212)$$

b) Metric system of units

$$W_2(0.248)(T_1 - T_2) = W_1(656.8 - T_3) + W_1(0.48)(T_2 - 100)$$

where

water addition rate	:	w_1 = lb/ton cl.	W_1 = kg/kg cl.
weight of exit gas (dry)	:	w_2 = lb/ton cl.	W_2 = kg/kg cl.
uncooled gas temperature	:	t_1 = °F	T_1 = °C
cooled exit gas temp.	:	t_2 = °F	T_2 = °C
water temperature	:	t_3 = °F	T_3 = °C

14.02 Kiln Feed Residence Time

The approximate time taken by the feed to travel the length of the kiln can be calculated by the following formulas:

a) English system of units

$$T = \frac{11.4x}{NdS}$$

b) Metric system of units

$$T = \frac{11.4L}{NDS}$$

where

T = travel time (min.)
x = length of kiln (ft)
L = length of kiln (m)
N = kiln speed (RPH)
d = kiln diameter (ft)
D = kiln diameter (m)
S = slope of kiln (ft/ft) or (m/m)

14.03 Kiln Slope Conversion

Slope	in./ft	m/m	Angular degrees
¼ in.	0.250	0.0208	1.192
5/16 in.	0.3125	0.0260	1.492
3/8 in.	0.375	0.0313	1.790
7/16 in.	0.4375	0.0365	2.088
½ in.	0.50	0.0417	2.386
9/16 in.	0.5625	0.0469	2.684
5/8 in.	0.625	0.0521	2.981
11/16 in.	0.6875	0.0573	3.279
¾ in.	0.750	0.0625	3.576
13/16 in.	0.8125	0.0677	3.873
7/8 in.	0.875	0.0729	4.170
15/16 in.	0.9375	0.0781	4.467
1 in.	1.000	0.0833	4.764

Slope is often expressed also as a percent of the kiln length.

$$\text{Percent slope} = \frac{(\text{ft/ft slope})L}{L}\,100 = \frac{(\text{m/m slope})L_m}{L_m}\,100$$

L = kiln length (ft)
L_m = kiln length (m)

14.04 Kiln Sulfur Balance

If a kiln performance study has been completed as outlined in Chapters 8 and 9, the necessary data below can be obtained from the data sheet given in these Chapters.

a) English system of units

Data needed:

A_s	=	percent sulfur, S, in fuel (as fired)	=	.....
W_A	=	lb fuel per ton of clinker (as fired)	=	.....
C_S	=	percent SO_3 in kiln feed	=	.....
W_{dF}	=	lb dry feed per ton of clinker	=	.....
H_S	=	percent SO_3 in clinker	=	.....
G_s	=	percent SO_3 in dust	=	.....
K	=	lb dust per ton of clinker (to be calculated)	=	.....
G	=	percent of collected dust that is returned to kiln (expressed as a decimal)	=	.....

SULFUR (SO_3) BALANCE

Input		lb/t cl	*Output*		lb/t cl*
Fuel :	$0.02497 A_s W_A$	.....	Clinker	: $20 H_s$	.....
Feed :	$(C_S/100) W_{dF}$	.....	Dust	: $(1 - G)(G_S/100)K$	.....
Dust :	$G(G_S/100)K$	.....	Exit gas :		.....
	Total:	.....		Total:	.....

*lb/ton clinker

Note: Sulfur in the exit gas is calculated by difference to make the two sides equal.

b) Metric units

A_s	=	percent sulfur, S, in fuel (as fired)	=	.....
W_A	=	kg fuel per kg clinker (as fired)	=	.....
C_s	=	percent SO_3 in kiln feed	=	.....
W_{dF}	=	kg dry feed per kg clinker	=	.....
H_s	=	percent SO_3 in clinker	=	.....
G_s	=	percent SO_3 in dust	=	.....
K	=	kg dust per kg clinker	=	.....
G	=	percent of collected dust that is returned to kiln (expressed as a decimal)	=	.....

SULFUR, SO_3 BALANCE

Input	kg/kg cl.	*Output*	kg/kg cl.
Fuel : $0.02497A_sW_A$	.....	Clinker : $0.01H_s$	.....
Feed : $0.01C_sW_{dF}$	.....	Dust : $(1-G)0.01G_sK$	.....
Dust : $0.01GG_sK$	.....	Exit gas :	.....
Total:	.....	Total:	.....

Note: Exit gas concentrations are calculated by difference to make the two sides equal in the total.

14.05 The Standard Coal Factor, Combustion Air Requirements

To determine the approximate combustion air needed to burn a given unit weight of coal, the formulas given below can be used when no ultimate coal analysis is available. The combustion air requirements include here 5 percent excess air.

a) English units

$$SCF = \frac{100 - a}{100} \; \frac{b}{12{,}600}$$

lb air/lb coal = 10.478 *SCF*

b) Metric units

$$SCF = \frac{100 - a}{100} \; \frac{B}{7000}$$

kg air/kg coal = 10.478 *SCF*

SCF = standard coal factor
a = percent moisture in coal (as fired)
b = heat value of coal (Btu/lb as fired)
B = heat value of coal (kcal/kg as fired)

14.06 Cooler Performance

English units	*Metric units*
$w = Ahd$	$w = Ahd$
$r = \frac{w}{2000\,W_{Cl}}\,60$	$r = \frac{w}{W_{Cl}}\,60$
$q = \frac{wc_c(t_1 - t_2)}{c_a(T_2 - T_1)}\,f$	$q = \frac{wc_c(t_1 - t_2)}{c_a(T_2 - T_1)}\,f$
$q_m = \frac{q}{r}$	$q_m = \frac{q}{r}$
$T_2 = \frac{wc_c(t_1 - t_2)}{qc_a} + T_1$	$T_2 = \frac{wc_c(t_1 - t_2)}{qc_a} + T_1$
$E = \frac{Q_c - Q_1}{Q_c}\,100$	$E = \frac{Q_c - Q_1}{Q_c}\,100$
$Q_c = 2000c_c(t_1 - 32)$	$Q_c = c_c t_1$

where

			English	Metric
A	=	cooler grate area	ft^2	m^2
c_c	=	mean specific heat of clinker		
c_a	=	mean specific heat of air		
d	=	clinker density	lb/ft^2	kg/m^3
E	=	thermal efficiency of cooler		
f	=	constant, 1.17		
h	=	clinker bed depth	ft	m
q	=	cooling air required	lb	kg
q_m	=	cooling air required	lb/min	kg/min
r	=	clinker residence time	min	min
t_1	=	temperature clinker in	°F	°C
t_2	=	temperature clinker out	°F	°C
T_1	=	temperature air in	°F	°C
T_2	=	temperature air out of clinker	°F	°C
Q_c	=	heat content of clinker, cooler in	Btu/ton	kcal/kg
Q_1	=	total heat losses in cooler	Btu/ton	kcal/kg
w	=	weight of clinker in cooler	lb	kg
W_{Cl}	=	kiln output	tph	kg/h

Note: The cooler heat losses, Q_1, can be obtained from the heat balance and radiation losses are estimated at $0.07Q_c$ for grate coolers, $0.10Q_c$ for planetary coolers. Q_1 must include all losses (clinker at discharge, cooler stack and radiation losses).

14.07 Combustion Air Required for Natural Gas Firing

In the absence of a complete analysis of the gas, the air requirements can be estimated from the following table. This table is based on natural gas with a heating value of 9345 $kcal/m^3$ (1050 Btu/ft^3).

O_2 *exit gas*	*Percent excess air*	CO_2 *combustion product percent by weight*	*Air required* lb *per million* Btu	kg *per* 1000 kcal
0.5	–	12.2	740	1.332
1.0	5.0	11.4	780	1.404
1.5	8.0	11.2	810	1.458
2.0	10.0	10.8	830	1.494
2.5	13.0	10.5	850	1.530
3.0	16.5	10.3	870	1.566
3.5	20.0	9.8	890	1.602

14.08 Products of Combustion on Natural Gas Firing

One standard cubic foot of gas, when burned, yields the following combustion products:

CO_2 = 0.1297 lb
H_2O = 0.1020 lb
N_2 = 0.6929 lb
O_2 = 0.0229 lb

Total = 0.9474 lb

One standard cubic meter of natural gas, when burned, yields the following combustion products:

CO_2	=	2.0778 kg
H_2O	=	1.6340 kg
N_2	=	11.1003 kg
O_2	=	0.3669 kg
Total	=	15.179 kg

14.09 Percent Loading of the Kiln

$$\text{Percent loading} = \frac{W_{Cl} W_{dF} T}{d_f 60 V}$$

where

		English	Metric
W_{Cl}	= kiln output	ton/h	kg/h
W_{dF}	= feed rate	lb/ton cl.	kg/kg cl.
T	= residence time (see **6.02**)	min	min
d_f	= bulk density of feed	lb/ft^3	kg/m^3
V	= internal kiln volume	ft^3	m^3

14.10 Cross-Sectional Loading of the Kiln

The formulas given here are applicable in either the metric or the English system of units.

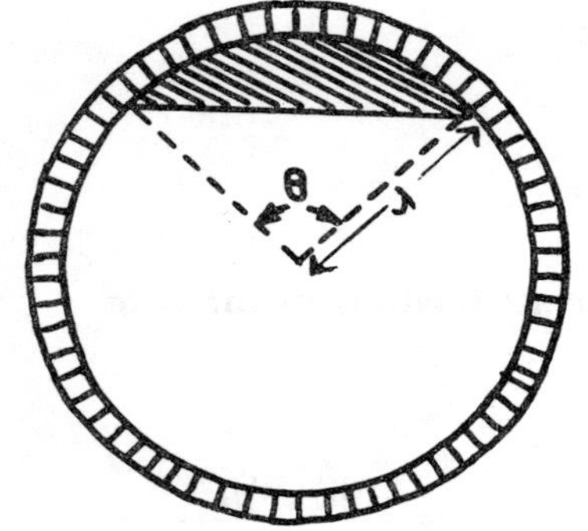

θ = radians
r = radius inside lining
A_1 = area occupied by feed

$$A_1 = \frac{r^2}{2}(\theta - \sin\theta)$$

$$\text{percent loading} = \frac{A_1}{\pi r^2}\,100$$

14.11 Flame Propagation Speed

For coal fired kilns, the primary air velocity should be at least twice as high as the flame propagation speed to prevent flash backs of the flame. Flame propagation is usually considerably lower than the velocity needed to convey coal dust by means of primary air into the kiln. Therefore, the minimum velocity necessary to convey coal without settling in ducts takes precedence over flame propagation speed when setting air flow rates or designing new burners (minimum velocity needed in ducts to prevent settling: 7000 ft/min, 35 m/s). Coal burners are usually designed to deliver a tip velocity of 9000–13,500 ft/min, 45 to 70 m/s.

ENGLISH UNITS

Primary air ft^3/lb coal	*Flame propagation* (ft/min)		
	30 percent VM 5 percent ash	30 percent VM 15 percent ash	10 percent VM* 7 percent ash
20	950	900	500
40	2250	1900	900
60	2800	2300	1300
80	2750	2180	1240
100	2550	1900	1050
120	2320	1670	930
140	2150	1500	840
160	2020	1350	750
180	1900	1250	690

*volatile matter

METRIC UNITS

Primary air m^3/kg coal	*Flame propagation* (m/s)		
	30 percent VM 5 percent ash	30 percent VM 15 percent ash	10 percent VM 7 percent ash
1	4.0	3.9	2.2
2	8.8	7.5	3.8
3	13.0	11.0	5.4
4	14.4	11.7	6.7
5	14.0	11.1	6.3
6	13.1	10.0	5.5
7	12.2	9.0	4.9
8	11.4	8.1	4.5
9	10.8	7.5	4.1
10	10.3	6.9	3.8
11	9.8	6.6	3.6

14.12 Kiln Drive Horsepower

ENGLISH UNITS

a) Friction horsepower

$$hp_f = \frac{W d_b d_t N F 0.0000092}{d_r}$$

b) Load horsepower

$$hp_l = (D \sin \theta)^3 NLK$$

c) Total kiln drive horsepower

$$hp_{total} = hp_f + hp_l$$

$$KW = (hp_{total})0.745$$

Note: Actual power input is usually 0.4 to 0.6 of hp_{total}.

W = total vertical load on roller shaft (lb)
d_b = diameter of roller shaft bearing (in.)
d_t = diamter of tire (in.)
d_r = diameter of rolls (in.)
N = kiln speed (rpm)
F = coefficient of friction (0.018 for oil, 0.06 for grease lubricated bearings)
L = kiln length (ft)
K = constant, 0.00076
D = kiln diameter (ft)

To calculate the load horsepower use the following approximate values for $\sin \theta$:

Percent kiln loading :	5	10	15
$\sin \theta$:	0.59	0.73	0.82

METRIC SYSTEM

a) Friction horsepower

$$hp_{mf} = \frac{0.000007873 W d_b d_t N F}{d_r}$$

b) Load horsepower

$$hp_{ml} = 0.086832(D \sin \theta)^3 NL$$

c) Total kiln drive horsepower

$$hp_{m,\text{total}} = hp_{mf} + hp_{ml}$$

$$kW = 0.7355 hp_{m,\text{total}}$$

For $\sin \theta$, use the following approximate values:

Percent kiln loading :	5	10	15
$\sin \theta$:	0.59	0.73	0.82

W = total vertical load on roller shaft (kg)
d_b = diameter of roller shaft bearing (cm)
d_t = diameter of tire (cm)
d_r = diameter of rolls (cm)
N = kiln speed (rpm)
F = constant, 0.018 for oil, 0.06 for grease lubricated bearing
L = kiln length (m)
D = kiln diameter (m)

14.13 Theoretical Exit Gas Composition, by Volume

In Chapter 9, Section **9.13**, the total weight of the exit gas components were calculated. In many studies, it is desirable to express this composition in terms of percent by volume. The following steps are taken to accomplish this. The formulas can be used for either English or metric units.

Step 1: Convert weights of each component into lb-moles or kg-moles as follows:

$$\frac{\text{weight of } CO_2}{44.01} = \ldots\ldots \text{ moles } CO_2$$

$$\frac{\text{weight of } N_2}{28.02} = \ldots\ldots \text{ moles } N_2$$

$$\frac{(0.23)(\text{weight excess air})}{32.0} = \ldots\ldots \text{ moles } O_2$$

$$\frac{\text{weight } SO_2}{64.06} = \ldots\ldots \text{ moles } SO_2$$

$$\frac{\text{weight } H_2O}{18.02} = \ldots\ldots \text{ moles } H_2O$$

Total: $= \ldots\ldots$ moles gas

Step 2: To obtain the percent by volume of any component, divide the moles of the component by the total moles of gas.

14.14 Conversion of Specific Heat Consumption into Annualized Costs

For U.S. currency:

$$\text{Dollars per year} = \frac{4.38 Q_e adc}{b}$$

For any other currency denomination:

$$\text{Monetary value/year} = \frac{8760 dCQ_m A}{B}$$

a = fuel costs (dollars/ton)
b = fuel heat value (Btu/lb)
c = kiln output (tph)
d = percent operating time (decimal)
A = cost per kg fuel
B = heat value fuel (kcal/kg)
C = kiln output (kg/h)
Q_e = specific heat consumption (Btu/t. clinker)
Q_m = specific heat consumption (kcal/kg cl.)

14.15 Theoretical Flame Temperature

(This formula applies only to oil or coal fired kilns)

English units

$$T = \frac{H_v}{0.06243Vs} + 32$$

Metric units

$$T = \frac{H_v}{Vs}$$

where

T = theoretical flame temperature (°F), (°C)
V = volume of combustion product (scfm/lb fuel), (std. m^3/kg fuel)
H_v = heating value of fuel (Btu/lb), (kcal/kg)
s = specific heat of combustion gas (use 0.38)

Note: "V" can be obtained by dividing the result of **8.11** or **9.11**, i.e., w_6 by $0.0847W_A$ when using English units and by dividing w_6 by 1.3569 when using metric units.

14.16 The "True" CO_2 Content in the Exit Gases

The true CO_2 content is the amount of carbon dioxide contained in the exit gases after a correction has been made to account for the effects of excess and deficiency of air present.

$$\text{"True" } CO_2 = \frac{100(CO_2 + CO)}{100 + 1.89CO - 4.78O_2}$$

14.17 Alkali Balance

DATA NEEDED

W_f	= feed rate (lb/lb cl.) or (kg/kg cl.)	=	.....
K_f	= percent K_2O in feed (expressed as a decimal)	=	.....
K_c	= percent K_2O in clinker (decimal)	=	.....
x	= percent of collected dust returned to kiln (decimal)	=	.....
N_f	= percent NaO_2 in feed (decimal)	=	.....
N_c	= percent NaO_2 in clinker (decimal)	=	.....
W_d	= total dust collected (lb/lb cl.) or (kg/kg cl.)	=	.....
K_d	= percent K_2O in dust	=	.....
W_e	= fuel rate (dry lb/lb cl.) or (kg/kg cl.)	=	.....
K_e	= percent K_2O in fuel (decimal)	=	.....
N_d	= percent NaO_2 in dust (decimal)	=	.....
N_e	= percent NaO_2 in fuel (decimal)	=	.....

ALKALI BALANCE

Input			*Output*		
Feed :	W_fK_f	=	Clinker :	K_c	= ....
	W_fN_f	=		N_c	= ....
Fuel :	W_eK_e	=	Dust :	$(1-x)W_dK_d$	= ....
	W_eN_e	=		$(1-x)W_dN_d$	= ....
Dust :	xW_dK_d	=	Exit gas :		= ....
	xW_dN_d	=			
	Total:	=		Total:	= ....

Exit gas concentrations are calculated by the difference to make the two sides equal.

14.18 Kiln Speed Conversions

$$r = \text{seconds per revolutions} = \frac{3600}{R}$$

$$R = \text{revolutions per hour} = \frac{3600}{r}$$

Peripheral speed

$$\text{in./s} = 0.00333R\pi D$$

$$= \frac{12\pi D}{r}$$

where

$$D = \text{kiln diameter (ft)}$$

$$\text{cm/s} = 0.02778R\pi D$$

$$= \frac{100\pi D}{r}$$

where

$$D = \text{kiln diameter (m)}$$

14.19 Power Audit on Kiln Equipment

Equipment	hp	kW	h/mo	kWh/mo
Kiln drives				
Cooler fans				
Cooler exhaust fan				
Cooler drives				
Clinker crusher				
Drag conveyors				
Clinker elevators				
Cooler dust collector				
Coal mill				
Coal conveyors and screws				
Nose cooling fan				
Shell cooling fans				
I.D. fan				
Precipitator				
Feed handling				
Dust handling (return)				
Dust handling (waste)				
F.K. pumps				
Compressors				
Others				
Total				

kWh/ton clinker: ....

kW = 0.7457 hp
h/mo = Total hours per month unit in operation
kWh/mo = (h/mo)(kW)

$$\text{kWh/ton clinker} = \frac{\text{kWh/mo}}{\text{monthly clinker production}}$$

14.20 Coating and Ring Formation

The graph shown here can serve as a guideline to indicate if a given clinker composition has the tendency to form heavy coatings and rings or if coating formation would be difficult. Clinkers that fall outside the shaded areas in their relationship between the silica ratio and the lime saturation factor tend to be either difficult or easy coating in nature depending on which side of the shaded area they are located. Clinker compositions that are located within the shaded area of this graph are considered acceptable from a coating formation viewpoint.

14.21 Relationship Silica Ratio vs. Saturation Factor

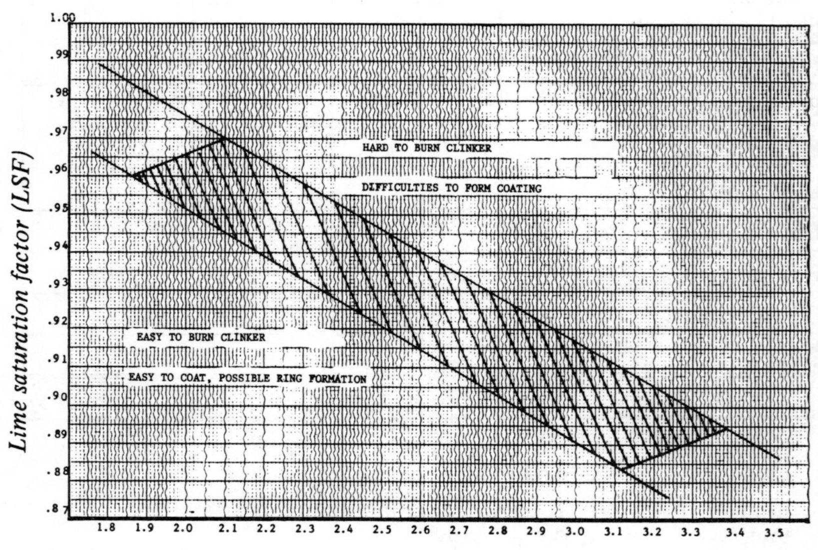

Silica ratio (S/R)

PROBLEMS AND SOLUTIONS

14.01 The exit gas temperature on a kiln is 285°C when a water spray rate of 0.19 kg water/kg clinker is used. Exit gas rate, dry, is 2.31 kg gas/kg clinker and the water temperature is 35°C. What is the exit gas temperature of the uncooled gases?

$$(2.31)(0.248)(T_1 - 285) = (0.19)(656.8 - 35) + (.19)(.48)(285 - 100)$$

$$T_1 - 285 = 236$$

$$T_1 = 521°\text{C} \quad (\textit{ans.})$$

14.02 A kiln has the following characteristics:
length: 125 m, diameter: 4.5 m, kiln speed: 68 rph, slope: 0.0417 m/m. What is the theoretical residence time of the feed in this kiln?

$$T = \frac{(11.4)(125)}{(68)(4.5)(0.0417)} = 112 \text{ min.} \quad (\textit{ans.})$$

14.03 Establish a sulfur balance (in the English system of units) for the wet process kiln given in example **1.01**. Use 2.3 percent SO_3 in kiln dust and a dust rate of 119.5 lb/t. clinker.

$A_s = 2.1$ $W_{dF} = 3220$ $k_1 = 119.5$

$W_A = 366$ $H_s = 0.2$ $G = 0.80$

$C_s = 0.1$ $G_s = 7.8$

SULFUR BALANCE

Input	lb SO_3/t. cl.	*Output*	lb SO_3/t
Fuel	19.19	Clinker	4.0
Feed	3.22	Dust waste	1.86
Dust return	7.46	Exit gas	14.03
Total	29.89	Total	29.89

14.05 What amount of combustion air is required per kg coal when the coal shows 3 percent moisture and a heat content of 6580 kcal/kg?

$$SCF = \left(\frac{100 - 3}{100}\right) \frac{6580}{7000} = 0.9118$$

$$\text{Combustion air required} = (10.478)(0.9118) = 9.55 \text{ kg/kg coal}$$

(*ans.*)

14.09 What is the percent loading of a kiln that shows:
feed consumption: 3220 lb/t. cl., residence time: 186 min, output rate: 30 tph, kiln volume: 38,000 ft^3, feed bulk density: 81 lb/ft^3?

$$\text{Percent loading} = \frac{(30)(3220)(186)}{(81)(60)(38{,}000)} = 0.097$$

$$= 9.7 \text{ percent} \quad (\textit{ans.})$$

14.15 What is the theoretical flame temperature on a kiln that shows 157.5 scfm combustion gas per pound of fuel and is fired with a coal of 12,050 Btu/lb?

$$T = \frac{12{,}050}{(0.06243)(157.5)(0.38)} = 3225°\text{F} \quad (\textit{ans.})$$

14.16 Given an exit gas analysis of:

CO_2	CO	O_2
25.6	0.1	0.8

What is the true CO_2 content in the gas?

$$\text{“True” } CO_2 = \frac{100(25.6 + 0.1)}{100 + (1.89)(0.1) - 4.78(0.8)} = 26.67 \quad (\textit{ans.})$$

Chapter 15

CHAIN SYSTEMS IN WET PROCESS KILNS

DATA NEEDED

			English	*Metric*
a	=	kiln output rate	. . . . . . tph	. . . . . . .kg/h
b	=	dry raw feed required	. . . . .t/t cl.	. . . .kg/kg cl.
c	=	slurry moisture (decimal)	. . . .	. . . .
d	=	total weight of chains	. . . . t	. . . . kg
e	=	effective kiln diameter (hanger to hanger)	. . . . ft	. . . . m
f	=	kiln diameter, inside lining	. . . . ft	. . . . m
g	=	total kiln length	. . . . ft	. . . . m
h	=	distance between hanger rings	. . . . ft	. . . . m
j	=	total chain surface area	. . . . . . .ft^2	. . . . m^2
l_1	=	total chain zone length	. . . . ft	. . . . m
l_2	=	evaporation zone length (feed inlet to zero feed moisture)	. . . . ft	. . . . m
l_3	=	length of chain	. . . . ft	. . . . m
m	=	attachment ring number (example: if chain strung from ring 1–4, $m = 4$)	. . . .	. . . .

		English	Metric
n	= attachment hole number (example: if chain departs hole 1 and arrives in hole 13, then n = 13)	....	....
p	= total number of holes/ring	....	....
t_{f_1}	= temperature feed chain inlet	.......°F	.........°C
t_{f_2}	= temperature feed chain outlet	.......°F	.........°C
t_{g_1}	= temperature gas, chain hot end	.......°F	.........°C
t_{g_2}	= temperature gas, kiln exit	.......°F	.........°C
x	= dust loss (percent in terms of fresh feed, expressed as a decimal)	....	....
z	= percent moisture in feed leaving the chains (decimal)	....	....
a_x	= threshold output rate for forced firing of the kiln	...... tph	......kg/h
A	= cross-sectional area, inside	.......ft^2	.... m^2
A_{ch}	= total chain surface area	.......ft^2	.... m^2
A_w	= wall surface area in chain section	.......ft^2	.... m^2
Q	= specific heat consumption	....MBtu/t	.....kcal/kg
k	= percent excess air in kiln (expressed as a decimal)	....	....

Note:

$$\Delta t_1 = t_{g_1} - t_{f_2}$$

$$\Delta t_2 = t_{g_2} - t_{f_1}$$

15.01 Chain Angle of Garland Hung Chains

There are three different methods commonly used to express the chain angle. To avoid confusion, we propose new terms for each in order to make a distinction between them.

a) The central chain angle

$$\theta = \frac{n-1}{p} 360$$

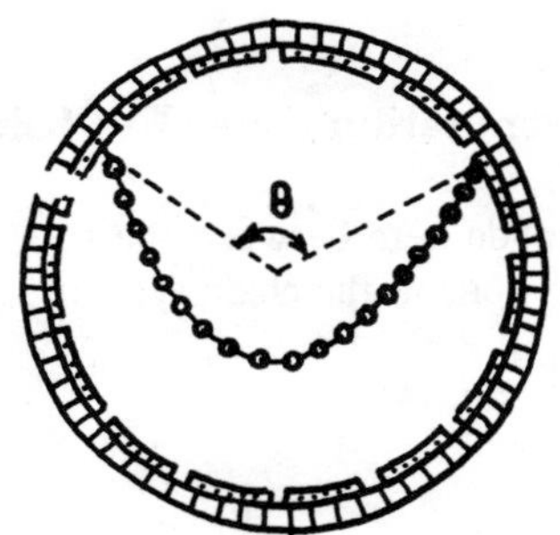

b) Exterior chain angle

Definition: The exterior angle formed between the straight line (connecting the two attachment holes of the chain) and the perpendicular to the kiln axis. This angle, ∡ , is calculated by using the result of (a) and proceeding in the following manner:

1. $A = e \sin \frac{1}{2}\theta$
2. $B = (m-1)h$
3. $\tan \phi = \frac{A}{B}$
4. exterior angle, ∡ $= \phi + 90$

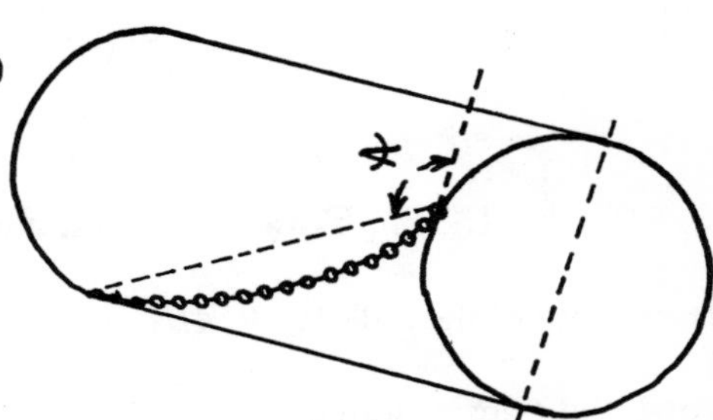

c) The chain length angle

The chain length expressed in degrees of the kiln circumference.

$$\phi = \frac{l_3}{\pi f} 360$$

15.02 Evaporation Rate (Wet Kiln)

A distinction must be made here between the total moisture given to the kiln and the evaporation done in the chain system.

a) Total evaporation in kiln

$$R_T = \frac{abc}{1 - c} \quad \text{(tons/h or kg/h)}$$

b) Evaporation in chain system

$$R_S = \frac{ab(c - z)}{1 - c + z} \quad \text{(tons/h or kg/h)}$$

c) Percent of evaporation done in chain system

$$Y = \frac{R_s}{R_T} 100$$

15.03 Total Heat Transfer Surface

The surface of the wall, $S_w = \pi f l_1$

Total heat transfer surface $= S_T = S_w + j$ (ft^2 or m^2)

15.04 Effective Heat Transfer Volume for Evaporation

$$V_T = \frac{f}{2}^2 l_1 \pi \text{ (ft}^3 \text{ or m}^3\text{)}$$

15.05 Chain Zone to Kiln Length Ratio

$$K = \frac{l_1}{g} 100$$

15.06 Length of Chain Systems

a) For garland hung systems

$$l_1 = h \text{ [(number of rows} - 1) + (m - 1)]$$

b) For curtain hung systems

$$l_1 = h\text{(number of rows)}$$

15.07 Chain Density

$$H = \frac{j}{\pi (0.5f)^2 l_1} \text{ (ft}^2\text{/ft}^3 \text{ or m}^2\text{/m}^3\text{)}$$

15.08 Heat Transfer Required in Chain System

Using English units:

$$Q_1 = \left\{(212 - t_{f_1}) + [0.48(t_{g_2} - 212)] + 970.3\right\} 0.002 R_S$$

(MBtu/h)

Using metric units:

$$Q_1 = [(100 - t_{f_1}) + 0.48(t_{g_2} - 100) + 539.1] R_S \text{ (kcal/h)}$$

$$Q_2 = ab0.002\sigma(t_{f_2} - t_{f_1})$$

Note $\sigma = 0.205$ for wet kilns
$\sigma = 0.260$ for dry kilns

Total heat transfer $= Q_T = Q_1 + Q_2$

15.09 Specific Chain System Performance Factors

a) Mass of chains per mass of water to be evaporated

$$F = \frac{d}{R_S}$$

English: (lb chains/h lb H_2O)
Metric: (kg chains/h kg H_2O)

b) Mass of chains per daily mass of clinker production

English system: $W = \dfrac{2000d}{24a}$ (lb chains/daily ton cl.)

Metric system: $W = \dfrac{41.67d}{a}$ (kg chains/daily metric ton cl.)

c) Specific evaporation per unit surface area

English system: $E_s = \dfrac{2000R_s}{S_T}$ (lb H_2O/h ft^2)

Metric system: $E_s = \dfrac{R_s}{S_T}$ (kg H_2O/h m^2)

d) Specific evaporation per unit kiln volume

English system: $$E_v = \frac{2000R_s}{V_T} \quad (\text{lb } H_2O/\text{h ft}^3)$$

Metric system: $$E_v = \frac{R_s}{V_T} \quad (\text{kg } H_2O/\text{h m}^3)$$

e) Specific heat transfer required per unit chain surface

English units: $$Q_{Sp} = \frac{Q_T \cdot 10^6}{j} \quad (\text{Btu/ft}^2)$$

Metric units: $$Q_{Sp} = \frac{Q_T}{j} \quad (\text{kcal/m}^2)$$

15.10 Chain System Design for Wet Process Kilns

A chain system design method is herein proposed that takes into account the amount of thermal work expected and the amount of heat made available in the system.

Step 1 Select the moisture in the feed leaving the chains.

$$z = \frac{\text{desired percent moisture}}{100} = \ldots\ldots$$

GUIDELINES

Feed	*Percent moisture after chains*
Poor nodulizing strength	6 – 10 percent
Medium nodulizing strength	3 – 7 percent
Good nodulizing strength	0 – 4 percent

Step 2 Select the appropriate output rate.

$$a_x = \frac{2186.2A}{893Q + 1100} = \text{.....} \text{ short tons/h}$$

This formula was developed by the author to show the theoretical threshold output rate for "forced firing" of the kiln. Forced firing is defined as that output rate at which excessive dust losses occur in the kiln as a result of the gas velocity exceeding 30 ft/s downstream of the chain system. Please note that only English systems of units are employed in this and all successive steps. For example, use only MBtu/short ton of clinker for "Q" in the above formula.

Step 3 Select the desired kiln exit gas temperature.

$$t_{g2} = \text{.....}\ ^{\circ}\text{F}$$

For optimum fuel efficiency, it is desirable to set this temperature as low as possible without causing condensation in the precipitator. If the temperature, at which condensation takes place, is known, the target should be set 50°F higher. For example, when it is known that condensation occurs at a kiln exit gas temperature of 350°F the target for t_{g_2} should be set at 400°F.

Step 4 Quantity of heat entering the chain system.

This empirical formula was developed by the author to obtain an approximate value for the heat entering the chain system.

$$Q_{in} = [0.0002137Q(1 + k) + 0.000281]\, t_{g_1} a + 0.00094(R_T - R_S)t_{g_1} = \text{.....} \text{ MBtu/h}$$

Find R_T and R_S in **15.02**.

Step 5 Quantity of heat leaving the chain system.

This formula was developed by the author to give an approximate value for the heat leaving the kiln at the feed end.

$$Q_{out} = [0.0002137Q(1+k) + 0.000281]\, t_{g_2} a + 0.00094 R_T t_{g_2}$$

$$Q_{out} = \ldots\ldots \text{MBtu/h}$$

Step 6 Heat supplied for thermal work in the chain system.

$$Q_\Delta = Q_{in} - Q_{out} = \ldots\ldots \text{MBtu/h}$$

Step 7 Ratio: heat supplied to heat transfer expected.

$$x = \frac{Q_\Delta}{Q_T} = \ldots\ldots$$

Find Q_T in **15.08.**

The value for "x" must be close to unity, i.e., 0.90–1.10. When this value is outside of this range, return to step 1 and adjust any of the variables such as z, a, c, e, t_{g_2}, or b to bring this value in line with the above given range.

Any adjustment in the variables causes the specific heat consumption of the kiln to change. To select the appropriate value for Q, the following guidelines can be used:

For t_{g_2} : a 10° increase is equivalent to $Q = +0.025$.
a 10° decrease is equivalent to $Q = -0.025$.

For a : a one ton increase is equivalent to $Q = -0.017$.
a one ton decrease is equivalent to $Q = +0.017$.

For c : a decrease of 0.01 in c is equivalent to $Q = -0.088$.
an increase of 0.01 in c is equivalent to $Q = +0.088$.

For b : a decrease of 0.10 in b is equivalent to $Q = -0.110$.
an increase of 0.10 in b is equivalent to $Q = +0.110$.

For e : a decrease of 0.01 in e is equivalent to $Q = -0.018$.
an increase of 0.01 in e is equivalent to $Q = +0.018$.

(If several of these factors are changed, the sum total changes in Q applies.)

Step 8 Chain surface area required.

Formula based on the logarithmic mean temperature differential in the chain system.

$$A_{ch} = \frac{Q_{\Delta}10^6}{f\dfrac{\Delta t_1 - \Delta t_2}{\ln\dfrac{\Delta t_1}{\Delta t_2}}} - AW = \ldots\ldots \text{ ft}^2$$

Guidelines for the value of f.

If z =	0	0.01	0.02	0.03	0.04	0.05	0.06
f =	2.6	2.9	3.7	3.9	4.0	4.1	4.3

z =	0.07	0.08	0.09	0.10	0.11	0.12	0.13
f =	4.8	5.5	6.1	6.6	6.7	6.8	6.8

(Note: When dust is insufflated, adjust f upward by using a "z" value 0.05 higher than actual, i.e., if $z = 0.01$, use $f = 4.3$ instead of $f = 2.9$.)

Step 9 Chain weight total.

$$W_{ch} = A_{ch}n = \ldots\ldots \text{ tons total}$$

For ¾ in. proof coil chains : $n = 0.003863$
For ¾ X 3 circle chains : $n = 0.003906$
For 1 in. circle chains : $n = 0.005310$

Step 10 Chain density.

$$D_{ch} = \frac{A_{ch}}{A\text{ (length of chain system)}} = \ldots\ldots \text{ ft}^2/\text{ft}^3$$

Guidelines:

Straight curtain hung : D_{ch} = max. 3.0
Spiral curtain hung : D_{ch} = max. 3.5
Garland hung : D_{ch} = max. 2.0

15.11 Kiln Chain Data–Round Links

Type steel	*Link size*	*Weight* lb/ft	*Surface* in.^{2}/ft	*Volume* in.^{3}/ft
MS	1 × 3 5/8	11.25	155.0	38.75
MS	1 × 3	12.0	162.7	40.68
SS	7/8 × 3	8.85	137.0	29.97
MS	¾ × 3	6.28	116.6	21.86
SS	¾ × 3	6.04	116.6	21.86
MS	¾ × 2¾	6.20	114.8	21.53
MS	5/8 × 2½	4.56	98.1	15.33

15.12 Kiln Chain Data–Proof Coil (Oval Links)

Type steel	*Link size* φ × inner width × inner length	*Weight* lb/ft	*Surface* in.^{2}/ft	*Volume* in.^{3}/ft
SS	19/32 × 1 11/32 × 3	8.12	143.76	21.34
MS	7/8 × 1 7/32 × 3 1/8	7.25	113.6	24.85
MS	¾ × (ribstile) 3	9.25	158.4	29.7
MS	¾ × 2¾	5.50	99.2	20.42
MS	¾ × 2 1/8	5.95	110.9	20.79
MS	11/16 × 2½	4.35	88.3	15.8
MS	5/8 × 2½	4.60	100.8	15.75
MS	5/8 × 2 3/16	3.50	78.8	12.31
MS	5/8 × 1 7/8	4.10	91.6	14.32

15.13 Chain Shackle Data

Size φ	*Weight* lb/piece	*Surface* in.^{2}/piece
¾ in.	2.2	41.5
7/8 in.	3.52	52.0
1 in.	4.8	68.5

15.14 Chain System Record Form

Plant: Kiln: Kiln Dimensions: Date:

Section	*A*	*B*	*C*	*D*	*E*	*F*	*Total*
Length of Section							
Type of Chain System							
Number of Rows							
Chains per Row							
Type of Chain Links							
Length per Chain							
Weight per unit length of chain							
Surface area per unit length chain							
TOTAL CHAIN WEIGHT							
TOTAL CHAIN LENGTH							
TOTAL CHAIN SURFACE AREA							
CHAIN DENSITY (ft^2/ft^3 or m^2/m^3)							

120 110 100 90 80 70 60 50 40 30 20 10

Distance between rings ------ # holes per ring ------ Rathole height ------ Distance feed end to chains ------

Kiln Operating Data		*Chain Data*		*Chain System Factors*
Kiln Output (tons per hour)	. . . . .	Total Tons Chains		R_T = ...
Raw Feed Required (t/t of clinker)	. . . . .	Garland Hanging Pattern		F = ...
Slurry Moisture (expressed as decimal)	. . . . .	From Ring......	To Ring......	W = ...
Spec. Fuel Consumption	. . . . .	From Hole......	To Hole......	E_s = ...
Rate of Dust Loss (percent)	. . . . .	Central Angle		E_v = ...
Chain Inlet Gas Temperature	. . . . .	Exterior Angle		
Feed End Temperature	. . . . .			
Material Moisture after Chains (percent)	. . . .	Date System Installed:		

Chapter 16

KILN REFRACTORY

16.01 Refractory Shapes

On the North American continent, rotary kiln blocks, arches, and wedges are the most common refractory shapes used to line rotary kilns. In countries using the metric system of units, VDZ and ISO shapes are used. The following data will familiarize the reader with the dimensional differences between these shapes. It is important to note that dimension "*a*", i.e., the back cord, is the face of the refractory that is in contact with the kiln shell. All shapes are installed so that the given dimension "*l*" forms a parallel line to the kiln axis. Dimension "*h*" indicates the lining thickness.

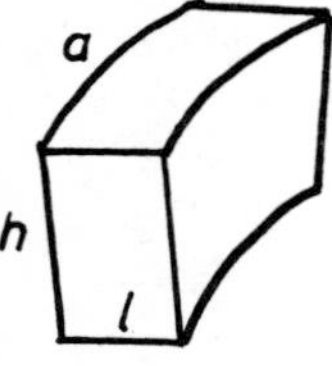

Rotary kiln block (RKB)

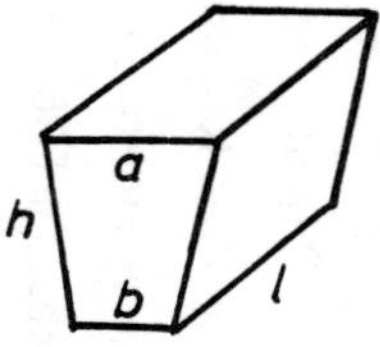

Arches

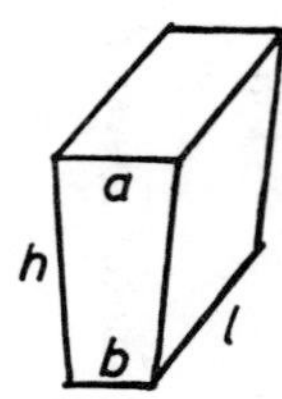

Wedges

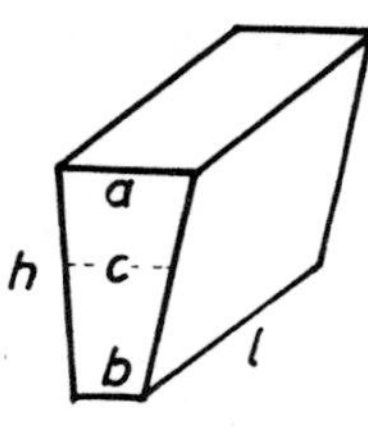

VDZ shapes

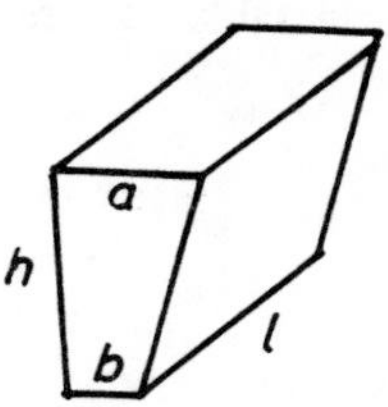

ISO shapes

a) Dimensions in the English system of units.

(dimensions given in inches)

	RKB			Arches			Wedges			VDZ	ISO π/3
a	9	9	12	4	3.5	3	4	3.5	3	var.	4
b	*v*	*v*	*v*	*v*	*v*	*v*	*v*	*v*	*v*	var.	var.
c	*v*	*v*	*v*	*v*	*v*	*v*	*v*	*v*	*v*	2.8–4	
l	4	4	4	9	9	9	6	6	6	7.8	7.8
h	6	9	9	6	6	6	9	9	9	var.	var.

Note: VDZ and ISO shapes are usually manufactured for 6 3/10 in., 7 in., 7 7/8 in., 8 7/8 in., and 9 4/5 in. lining thickness (*h*).

b) Dimensions in the metric system of units.

(dimensions given in millimeters)

	RKB			Arches			Wedges			VDZ	ISO π/3
a	229	229	305	102	89	76	102	89	76	var.	103
b	*v*	*v*	*v*	*v*	*v*	*v*	*v*	*v*	*v*	var.	*v*
c	*v*	*v*	*v*	*v*	*v*	*v*	*v*	*v*	*v*	72–102	
l	102	102	102	229	229	229	152	152	152	198	198
h	152	229	229	152	152	152	229	229	229	var.	var.

Note: VDZ and ISO shapes are usually manufactured for 160, 180, 200, 225, and 250 mm thick linings (*h*).

The following tables show typical VDZ shape dimensions in more details. These tables are reproduced with the permission of Refratechnik GmbH.

STANDARD SIZES FOR CEMENT ROTARY KILNS (Nonbasic Brick)

Symbol	Size	Dimensions in mm				Vol. dm^3	Kiln diameter (mm)
		a	b	h	l		
	A 216	107,5	89			3.130	1894
	A 416	104	95			3.130	3769
	A 616	102	97	160	198	3.130	6656
	AP 16	82	77			2.519	
	AP+16	92	87			2.835	
	A 218	110	89			3.560	1920
	A 418	105	95			3.560	3852
	A 618	103	97	180	198	3.560	6300
	AP 18	83	77			2.851	
	AP+18	93	87			3.208	
	A 220	112	89			3.980	1983
	A 420	106	95			3.980	3927
	A 620	104	97	200	198	3.980	6057
	AP 20	84	77			3.188	
	AP+20	94	87			3.584	
	A 222	114,5	89			4.500	2056
	A 422	107	95			4.500	4087
	A 622	105	97	225	198	4.500	6019
	AP 22	85	77			3.609	
	AP+22	95	87			4.054	
	A 225	114,5	89			5.040	2284
	A 425	108,5	95			5.040	4093
	A 625	106	97	250	198	5.040	6000
	AP 25	86	77			4.034	
	AP+25	96	87			4.529	
	A 230	103	72			5.198	2032
	A 430	103	87,5			5.658	4065
	A 630	103	92,5	300	198	5.806	6000
	AP 30	83	72,5			4.618	
	AP+30	93	82,5			5.212	

/ according VDZ WE 9 /

STANDARD SIZES FOR CEMENT ROTARY KILNS (Basic Bricks)

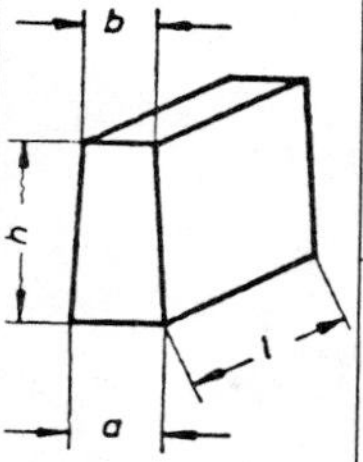

Symbol	Size	Dimensions in mm				Vol. dm³	*Kiln diameter* (mm)	
		a	b	h	l			
	B 216	78	65	160	198	2.265	1969	/ according VDZ WE 9 /
	B 416	75	68			2.265	3520	
	B 616	74	69			2.265	4864	
	BP 16	54	49			1.632		
	BP+16	64	59			1.948		
	B 218	78	65	180	198	2.548	2215	
	B 418	75	68			2.548	3960	
	B 618	74	69			2.548	5472	
	BP 18	54	49			1.835		
	BP+18	64	59			2.192		
	B 220	78	65	200	198	2.831	2462	
	B 420	75	68			2.831	4400	
	B 620	74	69			2.831	6080	
	BP 20	54	49			2.039		
	BP+20	64	59			2.435		
	B 222	78	65	225	198	3.185	2769	
	B 422	75	68			3.185	4950	
	B 622	74	69			3.185	6840	
	B 822	73	69			3.163	8438	
	BP 22	54	49			2.294		
	BP+22	64	59			2.740		
	B 225	78	65	250	198	3.540	3077	
	B 425	76	67			3.540	4333	
	B 625	74	69			3.540	7600	
	B 825	73	68,5			3.502	8333	
	BP 25	54	49			2.550		
	BP+25	64	59			3.045		
	B 427	76	66	275	198	3873	4290	
	B 627	74	68,5			3873	7600	
	BP 27	56	46			2859		
	BP+27	66	56			3349		

16.02 Number of Bricks Required per Ring

Outside the United States, it is customary to use two different shapes of bricks with different backcords (a) to complete a full circle of the kiln circumference. Experience has shown that this produces a superior fit of the refractory to the shell particularly when the kiln shell is slightly out of round.

a) for RKB, arches and wedges

for basic lining

$$n = \frac{12d\pi}{a + 0.059}$$

for aluminum brick

$$n = \frac{12d\pi}{a + 0.039}$$

n = number of bricks per ring
a = back cord (in.)
d = kiln diameter (ft)

b) for VDZ shapes

The attached tables, supplied by Refratechnik GmbH, show the number of bricks each required when a given kiln diameter is encircled with two different sizes of bricks. All of the brick shapes shown have a uniform dimension "*l*" of 198 mm.

NUMBER OF BRICKS REQUIRED FOR BASIC LINING (B-SHAPES)

Shapes	Kiln diameter (mm) 3000	3500	3600	3700	3800	3900	4000	4100	4200	4300	4400	4500	4600	4700	4800	4900	5000	5100	5200	5300	5400	5500
B218/B418	48/73	23/119	18/128	13/138	8/147	–/159																
B218/B618	66/55	53/89	50/96	48/103	45/110	42/117	39/124	37/131	34/138	31/145	29/152	26/159	23/166	21/173	18/180	15/187	13/194	10/201	7/208	–/210	–/224	–/228
B418/B618							158/5	147/21	136/36	126/50	115/66	104/81	93/96	83/111	72/126	61/141	51/156	40/171	29/186	18/201	8/216	–/228
B220/B420	70/50	45/96	40/106	35/115	30/124	25/134	20/143	15/157	10/161	5/171	–/180											
B220/B620	82/38	69/72	66/80	64/86	61/93	59/100	56/107	53/114	50/121	48/128	45/135	42/142	40/149	37/156	34/163	32/170	29/177	26/184	24/191	21/198	18/205	15/212
B420/B620												169/15	159/30	148/45	137/60	127/75	116/90	105/105	95/120	84/135	73/150	62/165
B222/B422	97/22	72/68	67/77	62/87	57/96	52/105	47/114	42/124	37/133	32/142	27/152	22/161	17/170	12/179	7/189	–/200						
B222/B622	103/16	89/51	86/58	84/65	81/72	78/79	75/86	73/93	71/99	68/106	65/114	63/120	60/127	57/134	55/141	52/148	49/155	47/162	44/169	41/176	39/183	36/190
B422/B622																	197/7	186/23	176/37	165/52	154/68	143/83
B225/B425		81/58	71/72	61/87	51/101	41/115	31/130	20/144	10/159	–/173												
B225/B625		108/31	105/38	103/45	100/52	97/59	95/66	92/72	90/79	87/86	84/93	82/100	79/107	76/114	74/121	71/127	68/135	66/142	63/148	61/155	58/162	55/169
B425/B/25											172/5	167/15	162/24	156/34	151/44	145/53	140/63	135/73	129/82	124/92	119/101	113/111

Example: How many bricks each are needed of the B 220 and B 420 shapes to complete a circle on a 3600 mm diameter kiln? Answer:
Answer: 40 pieces of shape B 220 and 106 pieces of shape B 420.

NUMBER OF BRICKS REQUIRED FOR NONBASIC LINING

Shapes	Kiln diameter (mm) 3000	3500	3600	3700	3800	3900	4000	4100	4200	4300	4400	4500	4600	4700	4800	4900	5000	5100	5200	5300	5400	5500
A281/A418	24/63	10/92	7/98	–/108	–/112	–/115																
A218/A618	40/47	34/68	33/72	31/77	30/82	29/86	28/89	26/94	25/99	24/103	23/107	21/112	20/116	19/120	18/124	17/129	15/133	14/137	13/141	11/147	10/151	9/155
A418/A618							106/11	101/19	97/27	92/35	87/43	83/50	78/58	73/66	68/74	64/82	59/89	54/97	49/105	45/113	40/121	35/129
A220/A420	26/60	12/90	9/96	6/102	–/111	–/114																
A220/A620	41/45	35/67	33/72	32/76	31/80	29/85	28/89	27/93	25/98	24/102	22/107	21/111	20/115	18/120	17/125	15/129	14/134	13/137	12/142	10/147	9/150	8/155
A420/A620							111/6	106/14	100/23	95/31	89/40	84/48	78/57	73/65	68/74	62/82	57/91	51/99	46/108	41/116	35/124	30/133
A222/A422	29/56	15/85	13/91	10/97	7/102	–/112	–/116															
A222/A622	42/43	35/65	34/70	33/74	31/78	30/82	29/87	27/92	26/96	25/100	23/105	22/109	20/114	19/118	17/123	16/127	15/131	13/136	12/141	11/145	9/150	8/153
A422/A622								114/5	108/14	102/23	96/32	91/40	85/49	79/58	73/67	67/76	61/85	55/94	50/103	44/112	38/121	32/129
A225/A425	37/47	20/79	17/85	13/92	10/98	7/104	–/114	–/117														
A225/A625	50/34	42/57	40/62	39/66	37/71	35/76	34/80	32/85	30/90	29/94	27/100	26/104	24/109	23/113	21/118	20/122	18/127	16/132	15/136	13/141	12/145	10/150
A425/A625									111/9	105/18	100/27	94/36	88/45	82/54	76/63	70/72	64/81	59/89	53/98	47/107	41/116	35/125

c) For I.S.O. shapes ($\pi/3$)

I.S.O. shapes have a uniform backcord of 103 mm. With an expansion insert of 1 mm, the cord length becomes 104 mm which is equivalent to $\pi/3$, explaining the reason for identifying these shapes by this nomenclature. With π a constant in the brick backcord and the circumference of the kiln shell, the calculation for the number of bricks required per circle becomes simple:

$$n = 1000D(0.0333) = 33.33D$$

where

$$D = \text{internal kiln shell diameter (m)}$$

Example: How many bricks are required in I.S.O. shapes to complete a circle on a 4.8 m diameter kiln?
Answer: (33.33)(4.8) = 160 pieces.

16.03 Number of Bricks Required per Unit Kiln Length

a) when dimension "l" is expressed in millimeters and kiln length in meters:

$N = \frac{\text{Brick}}{m}$	*RKB*	*Arches*	*Wedges*	*VDZ*	*ISO*
	$9.6n$	$4.3n$	$6.5n$	$5n$	$5n$

b) when dimension "l" is expressed in inches and kiln length in feet:

$N = \frac{\text{Brick}}{\text{ft}}$	*RKB*	*Arches*	*Wedges*	*VDZ*	*ISO*
	$3n$	$1.33n$	$2n$	$1.524n$	$1.524n$

n = number of bricks per ring (see **8.02**)

16.04 Kiln Diameter Conversion Table

ft	mm	ft	mm	ft	mm
8.84	2900	12.00	3937	15.54	5100
9.0	2953	12.19	4000	15.85	5200
9.14	3000	12.50	4101	16.00	5250
9.45	3100	12.80	4200	16.15	5300
9.5	3117	13.00	4265	16.46	5400
9.75	3200	13.11	4300	16.50	5414
10.00	3281	13.41	4400	16.75	5500
10.06	3300	13.50	4429	17.00	5577
10.36	3400	13.72	4500	17.07	5600
10.50	3445	14.00	4593	17.37	5700
10.66	3500	14.02	4600	17.50	5742
10.97	3600	14.32	4700	17.68	5800
11.00	3609	14.50	4757	18.00	5900
11.25	3691	14.63	4800	18.29	6000
11.27	3700	14.93	4900	18.50	6069
11.50	3773	15.00	4922	18.59	6100
11.58	3800	15.24	5000	18.90	6200
11.89	3900	15.50	5085	19.00	6234

PART III

GRINDING

Chapter 17

TECHNICAL INVESTIGATION OF GRINDING MILL

Formulas for a study of the grinding mill and circuit are given. To simplify the engineers task, all the formulas are presented in the form of work sheets that can be used in the course of the mill investigation. At the conclusion, a summary sheet is also given to compile all the significant results of this study. Data, formulas, and results can be presented either in English or metric system units by using the appropriate work sheets.

17.01 Technical Data of Grinding Circuit

Plant location: .... Mill: ....

Type of mill: ....

Manufactured by: .... Build: ....

Types of products ground: ....

....

....

Type of grinding circuit: ....

Separator

Type: .... Size: ....

Motor: ....

Mill drive

Type: ....

Motor: ....

	1. compartment		*2. compartment*		*3. compartment*	
diameter						
length						
volume						
BALL CHARGE	size	weight	size	weight	size	weight
Total						

Date of investigation: .... Tested by: ....

Chapter 18

GRINDING MILL INVESTIGATION (ENGLISH SYSTEM OF UNITS)

DATA NEEDED

B_f = specific surface Blaine, new feed .....

B_p = specific surface Blaine, finish product .....

C_1 = horizontal distance, liner to liner at ball charge surface, 1. compartment (ft) .....

C_2 = charge surface, 2. compartment (ft) .....

C_3 = charge surface, 3. compartment (ft) .....

$\cos \phi$ = mill motor power factor .....

D_1 = internal mill diameter, liner to liner (ft) .....

E = mill motor, applied voltage .....

f_p = percent passing, finish product (decimal) .....

f_r = percent passing, rejects (decimal) .....

f_s = percent passing, separator feed (decimal) .....

f_a = percent passing, new feed (decimal) .....

G_{HG} = Hardgrove grindability of material .....

G_{Ac} = Allis-Chalmers grinding index of material .....

H_1 = free vertical height, charge to liner, 1. compartment (ft) .....

H_2 = free vertical height, 2. compartment (ft) .....

H_3 = free vertical height, 3. compartment (ft) .....
I = mill motor amperes .....
L = internal length of mill, (ft) .....
N = mill speed (rpm) .....
P = mill horsepower .....
R = feed residence time in mill (min) .....
s = specific gravity of mill feed .....
t_m = temperature of feed, mill outlet, (°F) .....
t_c = temperature of separator product (°F) .....
U_1 = sieve size (microns) where 80 percent of the product passes .....
U_3 = sieve size (microns) where 80 percent of the new feed passes .....
V = volume of air passing through mill (ACFM) .....
W_1 = mill output (tph) .....

CALCULATIONS

18.01 Mill Critical Speed

$$c_s = \frac{76.63}{D_1^{1/2}} = \text{.....} \text{ rpm}$$

18.02 Percent of Critical Speed

$$s_o = \frac{N}{s_c} 100 = \text{.....} \text{ percent}$$

18.03 Ratio: Free Height to Mill Diameter

For 1. compartment: $r_1 = \dfrac{H_1}{D_1} =$ .....

For 2. compartment: $r_2 = \dfrac{H_2}{D_1} =$ .....

For 3. compartment: $r_3 = \dfrac{H_3}{D_1} =$ .....

18.04 Internal Volume of Mill

$$v_m = \pi\left(\frac{D_1}{2}\right)^2 L = \text{.....}\ \text{ft}^3$$

18.05 Percent Loading of the Mill

First, determine the central angle (θ) for each compartment by using the formula

$$\sin \tfrac{1}{2}\theta = \frac{C_{1,\,2,\,3,\ldots}}{D_1}$$

1. compartment θ_1 = ...
2. compartment θ_2 = ...
3. compartment θ_3 = ...

Second, determine the cross-sectional area of the mill

$$A = 0.25\,(D_1{}^2)\pi = \text{.....}\ \text{ft}^2$$

Third, determine the area occupied by the ball charge in each compartment

1. compartment: $A_1 = 0.008727\theta_1 r^2 - \dfrac{C_1(r - H_1)}{2} = \ldots\ldots$

2. compartment: $A_2 = 0.008727\theta_2 r^2 - \dfrac{C_2(r - H_2)}{2} = \ldots\ldots$

3. compartment: $A_3 = 0.008727\theta_3 r^2 - \dfrac{C_3(r - H_3)}{2} = \ldots\ldots$

Finally, determine the percent loading:

1. compartment: $l_1 = \dfrac{A_1}{A}$ $= \ldots\ldots$ percent

2. compartment: $l_2 = \dfrac{A_2}{A}$ $= \ldots\ldots$ percent

3. compartment: $l_3 = \dfrac{A_3}{A}$ $= \ldots\ldots$ percent

Average: $l_x = \ldots\ldots$ percent

Note: for the above calculations, $r = \dfrac{D_1}{2}$

18.06 Bulk Volume of Ball Charge

$$v_b = v_m l_x = \ldots\ldots \text{ ft}^3$$

(find v_m in **18.04** and l_x in **18.05**).

18.07 Weight of Ball Charge

In **17.01**, data on the mills ball charge is given. Since this data refers usually to the initial load of the mill, the following formula can be used to calculate the mill charge weight after the mill operated for a given length of time

$$w_b = 285 v_b = \ldots\ldots \text{ lb}$$

(find v_b in **18.06**).

18.08 Weight of Feed in Mill

$$w_f = \frac{2000 W_1 R}{60} = \ldots\ldots \text{ lb}$$

18.09 Steel to Clinker Ratio

$$r_2 = \frac{w_b}{w_f} = \ldots\ldots$$

(find w_b in **18.07** and w_f in **18.08**).

18.10 Bond's Laboratory Work Index

The work (kwh/t) required to reduce one short ton of a material from theoretical infinite size to 80 percent passing 100 microns.

The result applies to wet grinding in closed circuit.

$$w_i = \frac{435}{G_{HG}^{0.91}} = \ldots\ldots \text{ kWh/t}$$

or when the Allis-Chalmers grindability is used,

$$w_i = \frac{16}{G_{AC}^{0.82}} = \ldots\ldots \text{ kWh/t}$$

For dry grinding in a closed circuit, multiply w_i by 1.3333. For open circuit grinding, dry or wet, multiply w_i by 1.2.

When no Hardgrove or Allis Chalmers grindability ratings are available for a given material to be ground, use the Work Index guide lines given in Chapter 20.01.

18.11 Power Required

To grind a material from any feed size to any product size, the power required to perform this grinding work can be calculated by using the equation given by Bond for the Third Theory of Comminution:

$$w = \frac{10w_i}{(U_1)^{1/2}} - \frac{10w_i}{(U_3)^{1/2}} = \ldots\ldots \text{kwh/t}$$

When the product 80 percent passing size is less than 70 microns, the result in the above equation (w) has to be multiplied by the factor "f" which is determined as follows:

$$f = \frac{U_1 + 10.3}{1.145U_1} = \ldots\ldots$$

Thus,

$$fw = \ldots\ldots \text{kWh/t}$$

18.12 Mill Power

The kilowatts expanded in grinding, with an AC – 3 phase system is

$$\text{kW} = \frac{1.732EI \cos \phi}{1000} = \ldots\ldots$$

18.13 True Specific Power Demand of Grinding Mill

$$\text{kWh/t}_{\text{actual}} = \frac{\text{kW(hours operation)}}{\text{tons material ground}} = \ldots\ldots$$

(find kW in **18.12** or by actual field tests).

18.14 Mill Operating Efficiency

$$w_e = \frac{\text{kWh}/t_{\text{actual}}}{w} 100 = \ldots\ldots \text{ percent}$$

(find the numerator in **18.13** and the denominator in **18.11**).

18.15 Specific Surface Grinding Efficiency

$$s_e = \frac{w_1(B_P - B_F)}{\text{kW}} = \ldots\ldots \text{ specific surface per kWh}$$

(find kW in **18.12**).

18.16 Mill Size Ratio

$$r_3 = \frac{L}{D_1} = \ldots\ldots$$

18.17 Specific Mill Volume per Horsepower

$$s_v = \frac{v_m}{P} = \ldots\ldots \text{ ft}^3/\text{hp}$$

(find v_m in **18.04**).

18.18 Separator Load

For open circuit

$$w_3 = \frac{W_1(f_p - f_r)}{(f_s - f_r)} = \ldots\ldots \text{ tph}$$

For closed circuit

$$w_3 = \left(\frac{c_1}{100} + 1\right) W_1 = \ldots\ldots \text{ tph;}$$

18.19 Separator Efficiency

For all systems

$$s_e = \frac{f_p(f_s - f_r)}{f_s(f_p - f_r)} 100 = \ldots\ldots \text{ percent}$$

18.20 Circulating Load

For closed circuit, new feed, and mill product to separator

$$c_1 = \frac{f_p - f_a}{f_m - f_r} 100 = \ldots\ldots \text{ percent}$$

Note:

$$f_s = \frac{\left[f_r\left(\frac{c_1}{100}\right)\right] + f_p}{\frac{c_1}{100} + 1}$$

For closed circuit, new feed and rejects to mill,

if

$f_p > f_s$

then

$$c_1 = \frac{f_p - f_s}{f_s - f_r} 100 = \ldots\ldots \text{ percent}$$

if

$$f_s > f_p$$

then

$$c_1 = \frac{f_s - f_r}{f_r - f_s} 100 = \ldots\ldots \text{ percent}$$

18.21 Size of Grinding Balls Required

For steel balls, the largest ball diameter required for grinding is

$$B = \left[\frac{U_1 w_i}{200 c_s} \left(\frac{s}{D_1^{\frac{1}{2}}}\right)^{\frac{1}{2}}\right]^{\frac{1}{2}} = \ldots\ldots \text{ in.}$$

Another formula frequently used but not as accurate is:

$$B = \left(\frac{U_1}{K}\right)^{\frac{1}{2}} \left(\frac{s w_i}{c_s (D_1^{\frac{1}{2}})}\right)^{1/3} = \ldots\ldots \text{ in.}$$

where

K = 350 for wet grinding
K = 300 for dry grinding

(find c_s in **18.01**, w_i in **18.10**).

RESULTS OF GRINDING MILL STUDY (ENGLISH SYSTEM)

18.01	c_s	Mill critical speed	. . . . .rpm
18.02	s_o	Percent of critical speed	. . . . .percent
18.03	r_1, r_2, r_3	Ratio: free height/mill	1. . . . .
			2. . . . .
			3. . . . .
18.04	v_m	Internal mill volume	. . . . .ft^3
18.05	l_x	Percent loading of mill	. . . . .percent
18.06	v_b	Bulk volume, ball charge	. . . . .ft^3
18.07	w_b	Weight of ball charge	. . . . lb
18.08	w_f	Weight of feed in mill	. . . . lb
18.09	r_c	Steel to clinker ratio	. . . .
18.10	w_i	Laboratory work index	. . . .
18.11	w	Power for grinding	. . . . kWh/t
18.12	kW	Mill power expanded	. . . . kW
18.13	kWh/t	True power consumption	. . . . kWh/t
18.14	w_e	Mill operating efficiency	. . . . .percent
18.15	s_c	Specific surface grinding efficiency	. . . .cm^2/cm^3/kWh
18.16	r_s	Mill length/diameter ratio	. . . .
18.17	s_r	Specific volume per horsepower	. . . . ft^3/hp
18.18	w_3	Separator load	. . . . tph
18.19	s_e	Separator efficiency	. . . . .percent
18.20	c_l	Circulating load	. . . . .percent
18.21	B	Largest ball diameter required for grinding	. . . . .in.

PROBLEMS AND SOLUTIONS

18.01 A finish mill with a 4.0-m diameter rotates at 17.0 rpm. Determine the critical speed for this mill and calculate the percent critical speed this mill is operated at.

$$\text{Critical speed, } s_c = \frac{42.306}{4^{1/2}} = 21.15 \text{ rpm}$$

$$\text{Percent of critical speed} = \frac{17.0}{21.15} 100 = 80.4 \text{ percent} \quad (\textit{ans.})$$

18.07 Assume that a given compartment should have an optimum ball charge weight of 75 short tons. How many tons of balls have to be added to the compartment when the following dimensions are found:

D_1,	internal diameter,	= 13.25 ft
C_1,	horizontal distance, liner to liner at ball charge surface,	= 8.20 ft
H_1,	vertical distance, charge to liner,	= 11.75 ft
L_1,	length of compartment,	= 9.50 ft

Solution:

Step 1. Calculate the ratio: free height to mill diameter (see **18.03**):

$$r_1 = \frac{11.75}{13.25} = 0.8868$$

Step 2. Calculate the volume of the entire compartment:

$$v_m = \pi\left(\frac{13.25}{2}\right)^2 9.5 = 1309.92 \text{ ft}^3$$

Step 3. Determine the central angle (see **18.04**)

$$\sin \tfrac{1}{2}\theta = \frac{8.2}{13.25}, \qquad \theta = 76.5$$

The cross section of the mill is

$$A = 137.886 \text{ ft}^2$$

The area occupied by the ball charge

$$A_1 = 0.008727(76.5)(6.625)^2 - \frac{8.2(6.625 - 11.75)}{2}$$

$$= 50.315 \text{ ft}^2$$

The percent loading of the compartment

$$l_1 = \frac{50.315}{137.886} = 0.365$$

Note: the percent loading is expressed as a decimal

Step 4. Determine the bulk volume of the balls (see **18.06**)

$$v_b = (1309.92)(0.365) = 478.0 \text{ ft}^3$$

Finally, the weight of the charge is: (see **18.07**)

$$w_b = 285(478) = 136230 \text{ lb} = 68.115 \text{ tons}$$

Hence the make-up ball charge to be added to bring the load to the desired weight is

$$75.0 - 68.115 = 6.89 \text{ tons} \quad (ans.)$$

Chapter 19

GRINDING MILL INVESTIGATION (METRIC SYSTEM OF UNITS)

DATA NEEDED

B_f = specific surface Blaine, new feed .....

B_p = specific surface Blaine, finish product .....

C_1 = horizontal distance, liner to liner

1. compartment (m) .....

C_2 = 2. compartment (m) .....

C_3 = 3. compartment (m) .....

$\cos \phi$ = mill motor power factor .....

D_1 = internal diameter of mill, liner to liner (m) .....

E = mill motor applied voltage .....

f_p = percent passing, finish product (decimal) .....

f_r = percent passing, rejects (decimal) .....

f_s = percent passing, separator feed (decimal) .....

f_a = percent passing, new feed (decimal) .....

G_{HG} = Hardgrove grindability of material .....

H_1 = free vertical height, charge to liner

1. compartment (m) .....

H_2 = 2. compartment (m) .....

H_3 = 3. compartment (m) .....

I	= mill motor, amperes		.....
L	= internal length of mill (m)		.....
N	= mill speed (rpm)		.....
P	= mill horsepower		.....
R	= feed residence time in mill (min.)		.....
U_1	= sieve size (microns) where 80 percent of the products passes the sieve		.....
U_3	= sieve size (microns) where 80 percent of the new feed passes the sieve		.....
W_1	= mill output (product)	(metric t/h)	.....
s	= specific gravity of mill feed		.....

CALCULATIONS

19.01 Mill Critical Speed

$$c_s = \frac{42.306}{D_1^{\frac{1}{2}}} = \text{.....}\ \text{rpm}$$

19.02 Percent of Critical Speed

$$s_o = \frac{N}{s_c}\,100 = \text{.....}\ \text{percent}$$

19.03 Ratio: Free Height to Mill Diameter

For 1. compartment: $r_1 = \dfrac{H_1}{D_1} = \ldots\ldots$

For 2. compartment: $r_2 = \dfrac{H_2}{D_1} = \ldots\ldots$

For 3. compartment: $r_3 = \dfrac{H_3}{D_1} = \ldots\ldots$

19.04 Internal Volume of Mill

$$v_m = \pi \left(\frac{D_1}{2}\right)^2 L = \ldots\ldots \text{m}^3$$

19.05 Percent Loading of Mill

First, determine the central angle (θ) for each compartment by using the formula:

$$\sin \tfrac{1}{2}\theta = \frac{C_{1,2,3}}{D_1}$$

1. compartment $\theta_1 = \ldots\ldots$
2. compartment $\theta_2 = \ldots\ldots$
3. compartment $\theta_3 = \ldots\ldots$

Second, determine the cross-sectional area of the mill:

$$A = \pi \left(\frac{D_1}{2}\right)^2 = \ldots\ldots \text{m}^2$$

Third, determine area occupied by ball charge in each compartment:

1. compartment: $A_1 = 0.008727\theta_1 r^2 - \dfrac{C_1(r-H_1)}{2} = \ldots\ldots$

2. compartment: $A_2 = 0.008727\theta_2 r^2 - \dfrac{C_2(r-H_2)}{2} = \ldots\ldots$

3. compartment: $A_3 = 0.008727\theta_3 r^2 - \dfrac{C_3(r-H_3)}{2} = \ldots\ldots$

Finally, determine the percent loading:

1. compartment: $l_1 = \dfrac{A_1}{A}$ $= \ldots\ldots$ percent

2. compartment: $l_2 = \dfrac{A_2}{A}$ $= \ldots\ldots$ percent

3. compartment: $l_3 = \dfrac{A_3}{A}$ $= \ldots\ldots$ percent

Average: $l_x = \ldots\ldots$ percent

Note: for the above calculations, $r = \dfrac{D_1}{2}$

19.06 Bulk Volume of Ball Charge

$$v_b = v_m l_x = \ldots\ldots \text{m}^3$$

(find v_m in **19.04** and l_x in **19.05**).

19.07 Weight of the Ball Charge

In this chapter data on the mills ball charge is given. Since this data refers usually to the initial load of the mill, the following formula can be used to calculate the weight of the charge based on the bulk volume occupied

$$w_b = 4565 v_b = \ldots\ldots \text{ kg}$$

(find v_b in **19.06**).

19.08 Weight of Feed in Mill

$$w_f = \frac{1000 W_1 R}{60} = \ldots\ldots \text{ kg}$$

19.09 Steel to Clinker Ratio

$$r_2 = \frac{w_b}{w_f} = \ldots\ldots$$

(find w_b in **19.07** and w_f in **19.08**).

19.10 Bond's Laboratory Work Index

Definition: The work (kWh/t) required to reduce one metric ton of a material from theoretical infinite size to 80 percent passing 100 microns.

The result applies to wet grinding in closed circuit.

$$w_i = \frac{479.51}{G_{HG}^{0.91}} = \ldots\ldots \text{ kWh/t}$$

or

$$w_i = \frac{17.64}{G_{AC}^{0.82}}$$

For dry grinding in closed circuit, multiply w_i by 1.3333. For open circuit grinding, dry or wet, multiply w_i by 1.2. When no Hardgrove grindability ratings are available for a given material to be ground, use the Work Index guide lines given in Chapter 20, Section **20.01**.

19.11 Power Required

To grind a material from any feed size to any product size, the power required for grinding can be calculated from the equation given for Bond's Third Theory of Comminution

$$w = \frac{10w_i}{U_1^{1/2}} - \frac{10w_i}{U_3^{1/2}} \; 1.1023 = \ldots\ldots \text{ kWh/t}$$

When the product 80 percent passing size is less than 70 microns, the result (w) above must be multiplied by the factor "f"

$$f = \frac{U_1 + 10.3}{1.145U_1} = \ldots\ldots$$

Thus,

$$fw = \ldots\ldots \text{ kWh/t}$$

19.12 Mill Power

The kilowatts expanded in grinding, with an AC – 3 phase system is

$$\text{kW} = \frac{1.732EI\cos\phi}{1000} = \ldots\ldots$$

19.13 True Specific Power Demand of Grinding Mill

$$\text{kWh/t}_{\text{actual}} = \frac{\text{kW(hours operation)}}{\text{tons material ground}} = \ldots\ldots$$

(find kW in **19.12** or by actual field tests).

19.14 Mill Operating Efficiency

$$w_e = \frac{\text{kWh/t}_{\text{actual}}}{w} 100 = \ldots\ldots \text{ percent}$$

(find the numerator in **19.13** and the denominator in **19.11**).

19.15 Specific Surface Grinding Efficiency

$$s_e = \frac{W_1 (B_P - B_F)}{\text{kW}} = \ldots\ldots \text{specific surface per kWh}$$

(find kW in **19.12**).

19.16 Mill Size Ratio

$$r_3 = \frac{L}{D_1} = \ldots\ldots$$

19.17 Specific Mill Volume per Horsepower

$$s_v = \frac{v_m}{P} = \ldots\ldots \text{ m}^3\text{/hp}$$

(find v_m in **19.04**).

19.18 Separator Load

For open circuit

$$w_3 = \frac{W_T(f_p - f_r)}{(f_s - f_r)} = \ldots\ldots \text{ tph}$$

For closed circuit

$$w_3 = \left(\frac{c_l}{100}\right) + 1 \quad W_T = \;.....\; \text{tph}$$

19.19 Separator Efficiency

For all systems

$$s_e = \frac{f_p(f_s - f_r)}{f_s(f_p - f_r)} 100 = \;.....\; \text{percent}$$

19.20 Circulating Load

For closed circuit, new feed and mill product to separator

$$c_l = \frac{f_p - f_a}{f_m - f_r} 100 = \;.....\; \text{percent}$$

Note that

$$f_s = \frac{\left[f_r\left(\frac{c_l}{100}\right)\right] + f_p}{\frac{c_l}{100} + 1}$$

For closed circuit, new feed and rejects to mill

If

$f_p > f_s$

$$c_l = \frac{f_p - f_s}{f_s - f_r} 100 = \;.....\; \text{percent}$$

If

$$f_s > f_p$$

$$c_l = \frac{f_s - f_r}{f_r - f_s} 100 = \text{.....} \text{ percent}$$

19.21 Size of Grinding Balls Required

For steel balls, the largest ball diameter required is

$$B = 20.8545 \left[\frac{U_1 w_i}{200 c_s} \left(\frac{s}{D_1^{1/2}} \right)^{1/2} \right]^{1/2} = \text{.....mm}$$

Another formula frequently used but not as accurate is

$$B = 20.85 \left(\frac{U_1}{K} \right)^{1/2} \left(\frac{s w_i}{c_s D_1^{1/2}} \right)^{1/3} = \text{..... mm}$$

where

K = 350 for wet grinding
= 300 for dry grinding

(find c_s in **19.01**, w_i in **19.10**).

RESULTS OF GRINDING MILL STUDY (METRIC SYSTEM)

19.01	c_s	Mill critical speed	............rpm
19.02	s_o	Percent of critical speed	.........percent
19.03	r_1, r_2, r_3	Ratio: free height/mill	1. ....
			2. ....
			3. ....
19.04	v_m	Internal mill volume	.... m^3
19.05	l_x	Percent loading of mill	.........percent
19.06	v_b	Bulk volume, ball charge	.... m^3
19.07	w_b	Weight of ball charge	............kg
19.08	w_f	Weight of feed in mill	............kg
19.09	r_c	Steel to clinker ratio	....
19.10	w_i	Laboratory work index	....
19.11	w	Power for grinding	.... kWh/t
19.12	kW	Mill power expanded	.... kW
19.13	kWh/t	True power consumption	.... kWh/t
19.14	w_e	Mill operating efficiency	.........percent
19.15	s_c	Specific surface grinding efficiency	... cm^2/cm^3/kWh
19.16	r_s	Mill length/diameter ratio	....
19.17	s_r	Specific volume per horsepower	.... m^3/hp
19.18	w_3	Separator load	.... tph
19.19	s_e	Separator efficiency	.........percent
19.20	c_l	Circulating load	.........percent
19.21	B	Largest ball diameter required for grinding	............mm

PROBLEMS AND SOLUTIONS

19.09 What is the steel to clinker ratio of the following mill?

w_b	=	weight of ball charge	= 100,000 kg
w_1	=	mill output	= 42.5 tph
R	=	feed residence time	= 12.5 min

Solution:

First calculate the weight of feed in the mill

$$w_f = \frac{1000(42.5)(12.5)}{60} = 8854 \text{ kg}$$

Thus the steel to clinker ratio is

$$r_2 = \frac{100{,}000}{8854} = 11.3 \quad (ans.)$$

19.11 Clinker of 80 percent passing 3/8 in. has to be ground to a specific surface Blaine of 3200 cm^2/g. What is the power required (kWh/t) to do this grinding work?

Solution:

From the table given in **20.04**, 3200 Blaine = 80 percent $- 40.2\mu$ hence U_1 = 40.2. From the table given in **20.05**, 3/8 in. = 9510μ hence U_3 = 9510. From guidelines given in **20.01**, clinker, w_1 = 13.49.

Thus the power required is (see **19.11**)

$$w = \frac{10(13.49)}{40.2^{1/2}} - \frac{10(13.49)}{9510^{1/2}} = 19.89 \text{ kWh/t}$$

Now, since the product 80 percent passing size is less than 70μ

$$f = \frac{40.2 + 10.3}{1.145(40.2)} = 1.097$$

Thus the power required is

$$1.097 \times 19.89 = 21.82 \text{ kWh/t} \quad (\textit{ans.})$$

19.20 What is the circulating load when a given mill shows the following fineness passing the 325 sieve:

product, f_p = 92 percent
separator feed, f_s = 54 percent
rejects, f_r = 32 percent

Since

$$f_p > f_s,$$

$$c_l = \left(\frac{92 - 54}{54 - 32}\right)100 = 173 \text{ percent} \quad (\textit{ans.})$$

19.21 What is the largest ball diameter required (mm) for a mill whose critical speed is 21.15 rpm when clinker has to be ground to 80 percent passing 40μ sieve. The mill diameter is 4.0 m and the specific gravity of clinker equals 3.15.

$$B = 20.8545\left[\frac{(40)(14.87)}{200(21.15)}\left(\frac{3.15}{4^{1/2}}\right)^{1/2}\right]^{1/2} = 8.76 \text{ mm} \quad (\textit{ans.})$$

Chapter 20

USEFUL DATA FOR GRINDING MILL STUDY

20.01 Work Index for Various Materials

Clinker	3.09	13.49	14.87
Kiln feed	2.67	10.57	11.65
Clay	2.23	7.10	7.83
Coal	1.63	11.37	12.53
Gypsum rock	2.69	8.16	8.99
Limestone	2.68	10.18	11.22
Shale	2.58	16.40	18.07
Silica sand	2.65	16.46	18.14
Sand stone	2.68	11.53	12.71
Slag	2.93	15.76	17.37
Blast furnace slag	2.39	12.16	13.40

20.02 Size Distribution for a New Ball Charge in Mill

mm			100	90	80	60	50	40
	inches	4.5	4	3.5	3	2.5	2	1.5
	4.5	23						
100	4	31	23					
90	3.5	18	34	24				
80	3	15	21	38	31			
60	2.5	7	12	20.5	39	34		
50	2	3.8	6.5	11.5	19	43	40	
40	1.5	1.7	2.5	4.5	8	17	45	51
30	1	0.5	1	1.5	3	6	15	49
	Total percent	100	100	100	100	100	100	100

20.03 Grindability Factor

Mill output when other materials than clinker are ground in the same mill:

Rotary kiln clinker	1.00
Shaft kiln clinker	1.15-1.25
Blast furnace slag	0.55-1.10
Chalk	3.7
Clay	3.0-3.5
Marl	1.4
Limestone	1.2
Silica sand	0.6-0.7
Coal	0.8-1.6

20.04 Approximate 80 Percent Passing Size in Microns

The approximate value can be calculated from the specific surface Blaine or Wagner as follows:

$$\log x = 8.50 - 2.15 \log W$$

$$\log x = 2 \log \frac{20{,}300}{B} = \log \left(\frac{20{,}300}{B}\right)^2$$

Blaine	80 percent
2600	61.0
2800	52.6
3000	45.8
3100	42.9
3200	40.2
3300	37.8
3400	35.6
3500	33.6
3600	31.8
3700	30.1
3800	28.5
3900	27.1
4000	25.8
4100	24.5
4200	23.4
4400	21.3
4600	19.5
4800	17.9
5000	16.5
5500	13.6
6000	11.4

x = micron size, 80 percent passing
W = Wagner
B = Blaine

20.05 Screen Size Conversion to Micron Size

a) U.S. standard sieves

Screen	Micron
400	37
* 325	44
270	53
* 230	63
200	74
* 170	88
140	105
* 120	125
100	149
* 80	177
70	210
* 60	250
50	297
* 45	354
40	420
* 35	500
30	595
* 25	707
* 18	1000
* 10	2000
¼ in.	6350
5/16	8000
3/8	9510
7/16	11,200
½	12,700
5/8	16,000
¾	19,000
7/8	22,600
1.0	25,400
1¼	32,000
1½	38,100
2	50,800

*I.S.O. International Standard Sieves

b) I.S.O. Standard Internation Sieves

Sieve Number	*Aperture* (mm)	Microns
	22.6	
	16.0	
	11.2	
	8.0	
3½	5.66	
5	4.00	
7	2.83	
10	2.00	
14	1.41	
18	1.00	
25		707
35		500
45		354
60		250
80		177
120		125
170		88
230		63
325		44

20.06 Optimum SO_3 Content in Cement

Quantity required for 6-cube batch:

		Grams	
	mix 1	mix 2	mix 3
cement	940	930	920
graded standard sand	470	470	470
standard sand	470	470	470
gypsum	—	10	20

Calculation

$$G = \left(\frac{a}{a-b}\right).48 + d + 0.24$$

where

G = optimum SO_3
a = average strength both rounds of mix 2 – mix 1
b = average strength both rounds of mix 3 – mix 2
d = SO_3 in test cement

20.07 Calculations Related to Gypsum

Percent gypsum = (4.778)(percent combined H_2O)
Percent SO_3 in gypsum = (0.4651)(percent gypsum)
Percent SO_3 as anhydrite = percent SO_3 – percent SO_3 as gypsum
Percent anhydrite = (1.7003)(percent SO_3 as anhydrite)

20.08 Percent Gypsum Required for Desired SO_3 in Cement

$$x = \frac{a-c}{b} 100$$

where

x = percent gypsum to be added to clinker
a = desired percent SO_3 in cement
b = percent SO_3 in gypsum
c = percent SO_3 in clinker

20.09 Cement Fineness

$$s_v = pS$$

where

s_v = surface area (cm^2/cm^3)
S = specific surface (cm^2/g)
p = specific gravity

20.10 Heats of Hydration

C_3S = 120 cal/g
C_3A = 210 cal/g
C_4AF = 100 cal/g
CaO = 279 cal/g
MgO = 203 cal/g

20.11 Spray Cooling with Water

This formula applies to the cooling of gases as well as the cooling of solids.

a) English system

$$h = \frac{ws(T_1 - T_2)}{(212 - t_2) + 970f}$$

b) Metric system

$$h = \frac{ws(T_1 - T_2)}{(100 - t_2) + 538.9f}$$

where

h = water rate needed (lb/h), (kg/h)
w = material or gas rate (lb/h), (kg/h)
s = specific heat of gas or material
T_1 = initial gas or material temperature uncooled (°F), (°C)
T_2 = desired gas or material temperature after cooling (°F), (°C)

t_2 = water temperature (°F), (°C)
f = percent water evaporated (decimal)

Chapter 21

GRINDING AIDS AND CEMENT FINENESS

21.01 Grinding Aid Solutions

English units	*Metric units*
$c = 8.33a + db$	$C = A + DB$
$e = \dfrac{c}{a+b}$	$E = \dfrac{C}{A+B}$
$f = \dfrac{rb}{c}$	$F = \dfrac{rB}{C}$
$g = 0.01585v$	$G = \dfrac{60v}{1000}$
$h = gef$	$H = GEF$
$i = \dfrac{h}{m}$	$I = \dfrac{1000H}{M}$
$k = \dfrac{i}{20}$	$K = \dfrac{I}{10}$
a = water added (gal)	A = water added (ℓ)
b = grinding aid added (gal)	B = grinding aid added (ℓ)
c = weight of solution (lb)	C = weight of solution (kg)
d = density of grinding aid as received (lb/gal)	D = density of grinding aid as received (kg/ℓ)
e = density of solution (lb/gal)	E = density of solution (kg/ℓ)
f = percent solids in solution (expressed as decimal)	F = percent solids in solution (expressed as decimal)
r = percent solids in grinding aid as received (decimal)	
m = mill output rate (tph)	M = mill output rate (kg/h)
v = solution addition (cc/min)	
g = solution addition (gal/h)	G = solution addition (ℓ/h)
h = solids addition (lb/h)	H = solids addition (kg/h)
i = lb solids/ton cement	I = grams solids/kg cement
k = percent solids in cement	

21.02 Fineness of Portland Cement by Turbidimeter (Wagner)

This formula is only applicable to cement with specific gravity of 3.15.

$$S = \frac{38r(2.0 - \log I_{50})}{1.5 + 0.75 \log I_{7.5} + \log I_{10} + \log I_{15} + \log I_{45} - 9.5 \log I_{50}}$$

where

S = specific surface of sample (cm^2/g)
r = corrected weight percent of sample passing the No. 325 (45μ) sieve.
$I_{7.5}, I_{10}, \ldots, I_{50}$ = microammeter reading, μA, corresponding to particle diameters 7.5, 10, . . . , 50μ.

21.03 Table of Logarithms for Turbidimeter Microammeter Readings

This table has been developed to simplify the physical testers task in calculating the specific surface of a sample. The table should be copied and posted on or near the turbidimeter.

LOGARITHMS OF MIRCROAMMETER READINGS

		0.1	0.2	0.3	0.4	0.5	0.6	0.7	0.8	0.9
8.0	0.903	0.908	0.914	0.919	0.924	0.929	0.934	0.940	0.944	0.949
9.0	0.054	0.959	0.964	0.968	0.973	0.978	0.982	0.987	0.991	0.996
10.0	1.000	1.004	1.009	1.013	1.017	1.021	1.025	1.029	1.033	1.037
11.0	1.041	1.045	1.049	1.053	1.057	1.061	1.064	1.068	1.072	1.076
12.0	1.079	1.083	1.086	1.090	1.093	1.097	1.100	1.104	1.107	1.111
13.0	1.114	1.117	1.121	1.124	1.127	1.130	1.134	1.137	1.140	1.143
14.0	1.146	1.149	1.152	1.155	1.158	1.161	1.164	1.167	1.170	1.173
15.0	1.176	1.179	1.182	1.185	1.188	1.190	1.193	1.196	1.199	1.201
16.0	1.204	1.207	1.210	1.212	1.215	1.217	1.220	1.223	1.225	1.228
17.0	1.230	1.233	1.236	1.238	1.241	1.243	1.246	1.248	1.250	1.253
18.0	1.255	1.258	1.260	1.262	1.265	1.267	1.270	1.272	1.274	1.276
19.0	1.279	1.281	1.283	1.286	1.288	1.290	1.292	1.294	1.297	1.299
20.0	1.301	1.303	1.305	1.307	1.310	1.312	1.314	1.316	1.318	1.320
21.0	1.322	1.324	1.326	1.328	1.330	1.332	1.334	1.336	1.338	1.340
22.0	1.342	1.344	1.346	1.348	1.350	1.352	1.354	1.356	1.358	1.360
23.0	1.362	1.364	1.365	1.367	1.369	1.371	1.373	1.375	1.377	1.378
24.0	1.380	1.382	1.384	1.386	1.388	1.389	1.391	1.393	1.394	1.396
25.0	1.398	1.400	1.401	1.403	1.405	1.407	1.408	1.410	1.412	1.413
26.0	1.415	1.417	1.418	1.420	1.422	1.423	1.425	1.427	1.428	1.430
27.0	1.431	1.433	1.435	1.436	1.438	1.440	1.441	1.443	1.444	1.446
28.0	1.447	1.449	1.450	1.452	1.453	1.455	1.456	1.458	1.459	1.461

21.04 Particle Size Distribution

The turbidimeter test data can be used to calculate the particle size distribution of a cement sample. For a detailed description of the calculations refer to the Appendix of the specification ASTM C-115, Part 9. The worksheet below can be used in the computation of this particle size distribution.

	Percent by weight fraction	Percent by weight cummulative
$\log I_{45} - \log I_{50}$ = ... X 47.5 = ...	....	....
$\log I_{40} - \log I_{45}$ = ... X 42.5 = ...	....	....
$\log I_{35} - \log I_{40}$ = ... X 37.5 = ...	....	....
$\log I_{30} - \log I_{35}$ = ... X 32.5 = ...	....	....
$\log I_{25} - \log I_{30}$ = ... X 27.5 = ...	....	....
$\log I_{20} - \log I_{25}$ = ... X 22.5 = ...	....	....
$\log I_{15} - \log I_{20}$ = ... X 17.5 = ...	....	....
$\log I_{10} - \log I_{15}$ = ... X 12.5 = ...	....	....
$\log I_{7.5} - \log I_{10}$ = ... X 8.8 = ...	....	....
$2.00 - \log I_{7.5}$ = ... X 3.8 = ...	....	....

total x_t = ...

$$F = \frac{r}{\text{total } x} = \; .....$$

Fractional percent $= Fx_{1,2,\ldots}$

Note: r = percent passing the 325 mesh sieve (45μ)

PROBLEMS AND SOLUTIONS

21.03 Determine the specific surface and the particle size distribution of the cement sample given below. The microammeter readings from the turbidimeter are shown in the first column under "I".

Cement sample No:	. . . . .
Specific gravity:	3.15
Percent passing 325 sieve:	90.5

Surface determination				*Particle size distribution*			
Particle size	I	$\log I$		$\log I - \log I$	x	percent	cummulative
50	11.0		1.041	0.016 (47.5) =	0.760	7.4	90.5
45	11.4	1.057		0.015 (42.5) =	0.638	6.2	83.1
40	11.8	1.072		0.011 (37.5) =	0.413	4.0	76.9
35	12.1	1.083		0.034 (32.5) =	1.105	10.8	72.9
30	13.1	1.117		0.053 (27.5) =	1.458	14.2	62.1
25	14.8	1.170		0.023 (22.5) =	0.518	5.1	47.9
20	15.6	1.193		0.043 (17.5) =	0.753	7.3	42.8
15	17.2	1.236		0.063 (12.5) =	0.788	7.7	35.5
10	19.9	1.299		0.035 (8.8) =	0.308	3.0	27.8
7.5	21.6		1.334	0.666 (3.8) =	2.531	24.7	24.8
$0.75(\log I_{7.5})$		= 1.001		Total x_t	9.272		
		1.500		$F = \dfrac{90.5}{9.272} = 9.76$			
Subtotal:		11.728					
$-9.5 \log I_{50}$ =		−9.890					
		1.838					

$$S = \frac{38(90.5)(2.0 - 1.041)}{1.838}$$

$$S = 1794 = 1790 \text{ cm}^2/\text{g}$$

PART IV

ENGINEERING FORMULAS

Chapter 22

STEAM ENGINEERING

22.01 Latent Heat of Vaporization

This is the heat required to change 1 lb (English system) or 1 kg (metric system) of boiling water to steam.

Boiling temperature °F	°C	*Latent heat of vaporization* Btu/lb	kcal/kg
32	0	1075.8	597.7
50	10	1065.6	592.0
68	20	1055.5	586.4
86	30	1045.2	580.7
104	40	1034.9	574.9
122	50	1024.6	569.2
140	60	1014.1	563.4
158	70	1003.5	557.5
176	80	992.7	551.5
194	90	981.6	545.3
212	100	970.3	539.1
302	150	908.6	504.8
392	200	833.6	463.1
482	250	736.9	409.4
572	300	594.1	330.1
662	350	383.1	212.8

22.02 Saturated Steam Pressure

(Rankine's formula)

$$\log p = 6.1007 - \frac{2730}{T} + \frac{393{,}670}{T^2}$$

where

p = absolute pressure (psi + 14.7)
T = absolute temperature (F + 460)

Example: What is the absolute pressure of saturated steam at 245°F?

$$\log p = 6.1007 - \frac{2730}{705} + \frac{393{,}670}{705^2} = 1.43631$$

$$p = 27.31 \text{ psi (absolute)} \quad (ans.)$$

22.03 Enthalpy

This is the heat required to change the state of water or ice.

a) Enthalpy of liquid

to change 1 lb water from 32° to boil = 180 Btu's
to change 1 kg water from 0° C to boil = 100 kcal.

b) Enthalpy of vaporization (at atmospheric pressure)

to change 1 lb of water from boil to steam = 970.2 Btu's
to change 1 kg of water from boil to steam = 539.1 Btu's

c) Enthalpy of fusion

to change 1 lb of ice to water = 144 Btu's
to change 1 kg of ice to water = 80 Btu's

22.04 Superheated Steam

Saturated steam shows the same temperature as the water during evaporation. Superheated steam is defined as the condition where all the water has evaporated and the steam temperature has been raised.

$$v = 0.591 \frac{Tw}{p} - 0.135w$$

$$Q = 0.48[T - 10.27(P)^{1/4}] + 857.2$$

where

v = volume of steam (ft^3)
T = absolute temperature (°F + 460)
p = absolute pressure (psi + 14.7)
w = weight of steam (lb)
P = pressure of steam (lb/ft^2)
Q = Btu's required

Example: a) What volume does 1 lb of steam occupy at 14.6 $psi_{abs.}$ and 480°F?

$$v = 0.591 \frac{940 \times 1}{14.6} - (0.135 \times 1) = 37.91 \text{ ft}^3 \quad (ans.)$$

b) What amount of heat is required to produce 1 lb of steam at 450°F and 2100 lb/ft^2 pressure?

$$Q = 0.48[910 - 10.27\,(2100)^{1/4}] + 857.2 = 1260.6 \text{ Btu's} \quad (ans.)$$

22.05 Properties of Steam

a) English system of units

SATURATED STEAM

°F	p	v_g	h_g
32	.08854	3306	1075.8
40	.12170	2444	1079.3
50	.17811	1703.2	1083.7
60	.2563	1206.7	1088.0
70	.3631	867.9	1092.3
80	.5069	633.1	1096.6
90	.6982	468.0	1100.9
100	.9492	350.4	1105.2
120	1.6924	203.27	1113.7
140	2.8886	123.01	1122.0
160	4.741	77.29	1130.2
180	7.510	50.23	1138.1
200	11.526	33.64	1145.9
212	14.696	26.8	1150.4
240	24.969	16.323	1160.5
260	35.429	11.763	1167.3
280	49.203	8.645	1173.8
300	67.013	6.466	1179.7
320	89.66	4.91	1185.2
340	118.01	3.79	1190.1
360	153.04	2.96	1194.4
380	195.77	2.34	1198.1
400	247.31	1.86	1201.0
450	422.6	1.10	1204.6
500	680.8	0.6749	1201.7
550	1045.2	0.424	1190.0
600	1542.9	0.2668	1165.5

$$\text{density of steam} = \frac{1}{v_g} = \text{lb/ft}^3$$

p = absolute pressure, 14.7 + gage pressure (psi)
v_g = specific volume, ft^3/lb of steam
h_g = enthalpy of saturated steam, i.e., the heat required to raise water from 32°F and turn it into steam at the stated pressure.

b) Metric system of units

SATURATED STEAM

Temp. (°C)	p (atm.)	v_g (m^3/kg)	h_g (kcal)
0	0.00602	206.39	597.7
4	0.00828	152.57	599.6
10	0.01212	106.33	602.1
16	0.01744	75.33	604.4
21	0.0247	54.18	606.8
27	0.0345	39.52	609.2
32	0.0475	29.22	611.6
38	0.0646	21.87	614.0
49	0.1152	12.69	618.7
60	0.1966	7.68	623.3
71	0.3226	4.83	627.9
82	0.5110	3.14	632.3
93	0.7843	2.10	636.6
100	1.000	1.67	639.1
116	1.699	1.02	644.7
127	2.4108	0.73	648.5
138	3.348	0.54	652.1
149	4.560	0.404	655.4
160	6.10	0.307	658.4
171	8.03	0.237	661.2
182	10.41	0.185	663.6
193	13.32	0.146	665.6
203	16.83	0.116	667.2
232	28.76	0.069	669.2

Chapter 23

ELECTRICAL ENGINEERING

23.01 The Basic Formulas

$$P = \frac{E^2}{R} = I^2R = EI$$

$$R = \frac{E^2}{P} = \frac{P}{I^2} = \frac{E}{I}$$

$$E = (PR)^{1/2} = \frac{P}{I} = IR$$

$$I = \frac{P}{E} = \frac{P^{1/2}}{R^{1/2}} = \frac{E}{R}$$

I = current (amps)
E = electromotive force (volts)
R = resistance (ohms)
P = power (watts)

23.02 Direct Current Circuits

a) Series circuits (DC)

$$R_t = R_1 + R_2 + \ldots + R_n$$

$$E_1 = I_t R_1$$

$$E_2 = I_t R_2$$

$$E_t = E_1 + E_2 + \ldots + E_n$$

$$P_1 = I_t^2 R_1$$

$$P_2 = I_t^2 R_2$$

$$P_t = I_t + E_t$$

$$= P_1 + P_2 + \ldots + P_n$$

R_t = total resistance
E_t = total EM force
P_t = total power

b) Parallel circuits (DC)

$$I_1 = \frac{E_t}{R_1} \quad \text{and} \quad I_2 = \frac{E_t}{R_2}$$

$$I_t = I_1 + I_2 + \ldots + I_n$$

$$R_t = \frac{R_1 R_2 \ldots R_n}{R_1 + R_2 + \ldots + R_n}$$

I_t = total current

23.03 Alternating Current (AC)

a) Characteristic values in a cycle

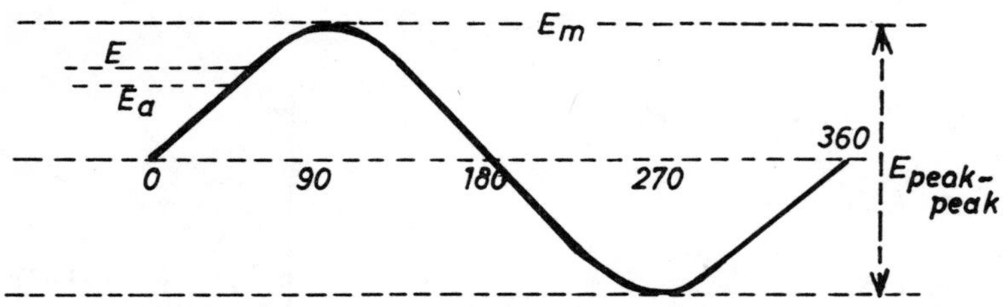

E_m = maximum voltage = $1.414E$ = $1.572E_a$ = $0.5E_{peak\text{-}peak}$	E = effective voltage = $0.707E_m$ = $1.11E_a$ = $0.35E_{peak\text{-}peak}$
E_a = average voltage = $0.636E_m$ = $0.9E$ = $0.318E_{peak\text{-}peak}$	$E_{peak\text{-}peak}$ = peak to peak voltage = $2.0E_m$ = $2.828E$ = $3.14E_a$

Notes: When converting AC to DC voltage, the average voltage, E_a, is used in the calculation. The characteristic curve for current is the same as for voltage, i.e., replace the above signs E with I to obtain the values for current. For AC circuit calculations, effective voltage and effective current are used. Most AC measuring instruments show effective voltages and currents.

b) Frequency of an AC cycle (hertz, cps)

Cycle frequencies are expressed in terms of cycles per second (cps) and hertz (hz) both of which have the same meaning. In other words, if an AC has 60 hz or 60 cps, one complete cycle takes 1/60 of a second.

c) Inductance

Unit: henry

$$x_L = 2\pi f L$$

where

x_L = inductive reactance (ohms)
f = frequency (hz)
L = inductance (henry)

d) Capacitance

Unit: farad

$$C = \frac{Q}{E}$$

$$C_1 = 0.2248 \frac{KA}{(1.0 \times 10^6)d}$$

where

C = capacitance (microfarads)
K = dielectric constant
A = effective area of plates (in.^{2})
d = thickness of dielectric (in.)
C_1 = capacitance (farads)
Q = charge (coulombs)
E = electromotive force (volts)

e) Generator

$$V_t = E - \text{internal } RI$$

$$W_1 = \frac{1}{\text{efficiency}} V_t I$$

f) Motor

$$V_t = E + \text{internal } RI$$

$$W = (\text{efficiency}) V_t I$$

g) Impedance

Definition: The total opposition to an alternating current including resistance and reactance.

$$Z = \frac{E}{I}$$

For an "RL" (resistance-reactance) circuit:

$$Z = R^2 + x_L{}^2$$

For an "RC" (resistance-reactance capacitance) circuit:

$$Z = R^2 + x_c{}^2$$

h) Capacitive reactance

$$x_c = \frac{1}{2\pi f C_1}$$

C_1 = capacitance (farads)
E = electromotive force (volts)
f = frequency (hz)
I = current (amps)
R = resistance (ohms)
x_L = inductive reactance (ohms)
x_c = capacitive reactance (ohms)
V_t = terminal voltage
W = mechanical power output
W_1 = mechanical power consumption
Z = impedance (ohms)

23.04 Useful Electrical Formulas

a) Direct current

$$A = \frac{hp\,746}{V(eff)}$$

$$kW = \frac{VA}{1000}$$

$$A = \frac{1000\,kW}{V}$$

$$hp = \frac{VA(eff)}{746}$$

b) A-C current–single phase

$$A = \frac{hp\,746}{V(eff)PF}$$

$$A = \frac{(kV - a)\,1000}{V}$$

$$kV - a = \frac{AV}{1000}$$

$$A = \frac{kW\,1000}{(V)PF}$$

$$kW = \frac{AVPF}{1000}$$

$$hp = \frac{AV(eff)PF}{746}$$

c) A-C current–two phase–four wire

$$A = \frac{746hp}{2V(eff)PF}$$

$$A = \frac{kW1000}{2VPF}$$

$$A = \frac{(kV - a)1000}{2V}$$

$$kW = \frac{VA2PF}{1000}$$

$$hp = \frac{VA2(eff)PF}{746}$$

$$kV - a = \frac{2VA}{1000}$$

d) A-C current–three phase

$$A = \frac{746hp}{1.73V(eff)PF}$$

$$A = \frac{1000(kV - a)}{1.73V}$$

$$kV - a = \frac{1.73VA}{1000}$$

$$A = \frac{kW1000}{1.73VPF}$$

$$\text{kW} = \frac{1.73\text{VA}PF}{1000}$$

$$\text{hp} = \frac{\text{VA(eff)}1.73PF}{746}$$

Note: PF = power factor

PROBLEMS AND SOLUTIONS

a) In an AC circuit, the maximum current, I_m, is given as 95 A. What is the effective amperage on this unit?

Solution: from **23.02 (a)**,

$$I = 0.707\, I_m$$

hence

$$I = (0.707)(95) = 67.2 \text{ A} \quad (\textit{ans.})$$

b) An AC–three phase motor reads a current of 75 A, an effective voltage of 350 and a power factor of 0.93. What is the power usage of this motor?

Solution: from **23.03 (d)**

$$\text{kW} = \frac{(350)(75)(.93)(1.73)}{1000} = 42.2 \text{ kW} \quad (\textit{ans.})$$

c) What is the power usage on a DC-motor that shows 350 V and 75 A?

Solution: from **23.03 (a)**

$$\text{kW} = \frac{(350)(75)}{1000} = 26.3 \text{ kW} \quad (\textit{ans.})$$

d) The terminal voltage on an AC motor is 220 V, the current is 1.33 A and the mechanical power output is stated as 175 W. What is the efficiency of this motor?

Solution: from **23.03 (f)**

$$\text{efficiency} = \frac{175}{(220)(1.33)} = 0.598 = 0.60 \quad (\textit{ans.})$$

Chapter 24

FAN ENGINEERING

24.01 Fan Laws

$\frac{Q_1}{Q_2} = \frac{n_1}{n_2}$	$n_2 = n_1 \frac{Q_2}{Q_1}$	$Q_2 = Q_1 \frac{n_2}{n_1}$
$\frac{Q_1}{Q_2} = \left(\frac{p_1}{p_2}\right)^{1/2}$	$p_2 = p_1\left(\frac{Q_2}{Q_1}\right)^2$	$Q_2 = Q_1\left(\frac{p_2}{p_1}\right)^{1/2}$
$\frac{Q_1}{Q_2} = \left(\frac{h_1}{h_2}\right)^{1/3}$	$h_2 = h_1\left(\frac{Q_2}{Q_1}\right)^3$	$Q_2 = Q_1\left(\frac{h_2}{h_1}\right)^{1/3}$
$\frac{n_2}{n_1} = \left(\frac{p_2}{p_1}\right)^{1/2}$	$n_2 = n_1\left(\frac{p_2}{p_1}\right)^{1/2}$	$p_2 = p_1\left(\frac{n_2}{n_1}\right)^2$
$\frac{n_2}{n_1} = \left(\frac{h_2}{h_1}\right)^{1/3}$	$n_2 = n_1\left(\frac{h_2}{h_1}\right)^{1/3}$	$h_2 = h_1\left(\frac{n_2}{n_1}\right)^3$

When n and Q are constant:

$\frac{h_2}{h_1} = \frac{t_1}{t_2}$	$h_2 = h_1 \frac{t_1}{t_2}$
$\frac{p_2}{p_1} = \frac{t_1}{t_2}$	$p_2 = p_1 \frac{t_1}{t_2}$
$\frac{h_2}{h_1} = \frac{d_2}{d_1}$	$h_2 = h_1 \frac{d_2}{d_1}$
$\frac{p_2}{p_1} = \frac{d_2}{d_1}$	$p_2 = p_1 \frac{d_2}{d_1}$

Q = flow rate
p = fan static pressure
n = fan speed
h = fan horsepower
t = absolute temp. (F + 460)
d = air density

24.02 Total Efficiency of a Fan

$$\text{eff}_{total} = \frac{0.000157Q(p_{total})}{h}$$

Q = flow rate
h = horsepower

(Note: $p_{total} = p_{static} + p_{velocity}$)

24.03 Static Efficiency of a Fan

$$\text{eff}_{static} = \frac{0.000157Q(p_{static})}{h}$$

24.04 Air Horsepower

$$\text{air hp} = \frac{144Q(p_2 - p_1)}{33{,}000}$$

24.05 Shaft Horsepower

$$\text{shaft hp} = \frac{\text{air horsepower}}{\text{efficiency}}$$

24.06 Similar Fans

For fans operating at the same speed and handling the same gas:

$$\frac{Q_1}{Q_2} = \left(\frac{D_1}{D_2}\right)^2$$

$$\frac{h_1}{h_2} = \left(\frac{D_1}{D_2}\right)^2$$

24.07 Fan Static Pressure

$$p_{static} = p_2 + p_1$$

p_1 = inlet pressure

p_2 = outlet pressure

D = diameter

PROBLEMS AND SOLUTIONS

For all problems, given a fan with the following operating characteristics: flow rate–14,500, horsepower–25, fan speed–940, static pressure–12.0, density of air–0.071.

a) What flowrate is obtained when the horsepower on this fan is increased to 40 hp?

$$Q_2 = 14{,}500\,(40/25)^{1/3} = 16{,}959 = 17{,}000 \quad (\textit{ans.})$$

b) What fan speed is needed with the same motor to obtain a flowrate of 16,000?

$$n_2 = 940\ (16{,}000/14{,}500) = 1037 = 1040 \quad (\textit{ans.})$$

c) What is the fan static pressure when the fan speed is being increased to 1050?

$$p_2 = 12.0\,(1050/940)^2 = 14.97 = 15.0 \quad (\textit{ans.})$$

d) What percent less fan horsepower is needed to obtain the same flow rate when the density of the air is 0.063?

$$h_2 = 25(0.063/0.071) = 22.18$$

and

$$[(1.0 - (22.18/25.0)]\,100 = 11.3 \text{ percent} \quad (\textit{ans.})$$

Chapter 25

FLUID FLOW

25.01 Viscosity

This is defined as the readiness at which a fluid flows when acted upon by an external force.

Units:

Metric system of units

μ = centipoise = 0.01 poise = 0.01 dyne
poise = g/cm/s
dyne = dyne s/cm^2

English system of units

μ_e = lb/ft/s

μ_{1e} = slug/ft/s

25.02 Kinematic Viscosity

Kinematic viscosity is expressed in centistokes units

$$\nu_k = \frac{\mu}{p}$$

$$\nu_k = \frac{\mu}{s}$$

where

p = g/cm^3
s = specific gravity
μ = centipoise

25.03 Specific Weight

This is often also referred to as the weight density and it represents the weight of a fluid per unit volume. In the metric system the units g/cm^3 are most often used. The English system uses primarily lb/ft^3 to express the weight density.

25.04 Specific Volume

This is the reciprocal of the specific weight.

$$\nu_s = \frac{1}{g/cm^3}$$

$$\nu_s = \frac{1}{lb/ft^3}$$

where

$$\nu_s = \text{specific volume (cm}^3\text{/g or lb/ft}^3\text{)}$$

25.05 Specific Gravity

$$s = \frac{p_1}{p_w}$$

where

p_1 = specific weight of liquid at stated temperature

p_w = specific weight of water at standard temperature

$$s = \frac{141.5}{131.5 + {}^\circ\text{API}}$$

$$s = \frac{140}{130 + {}^\circ\text{Baume}}$$

(when the liquid is lighter than water).

$$s = \frac{145}{145 - {}^\circ\text{Baume}}$$

(when the liquid is heavier than water).

25.06 Mean Fluid Velocity

$$V = \frac{r}{A}$$

$$V = \frac{R}{Ap}$$

$$V = \frac{RV_e}{A}$$

$$V = 183.3\frac{r}{d^2}$$

(when r and d are in English units).

$$V = 12732.4\frac{r}{d^2}$$

(when r and d are in metric units).

$$V = 0.2122\frac{Q}{d^2}$$

(when Q and d are in metric units).

$$V = 0.408\frac{Q}{d^2}$$

(when Q and d are in English units).

V = mean velocity (ft/s or m/s)
r = flow rate (ft^3/s or m^3/s)
A = area of pipe (ft^2 or m^2)
R = flow rate (lb/s or kg/s)
p = specific weight (lb/ft^3 or kg/m^3)
V_e = specific volume (ft^3/lb or m^3/kg)
d = diameter of pipe (in. or cm)
Q = quantity (gal/min or liter/min)

25.07 Barometric Pressure

This is the atmospheric pressure above zero absolute. Barometric pressure is always positive.

25.08 Atmospheric Pressure

Standard conditions (0°C or 32°F at sea level)

atmospheric pressure = 14.7 lb/in.^{2}
atmospheric pressure = 29.90 in. Hg
atmospheric pressure = 760 mm Hg
atmospheric pressure = 101.22 kPa

Note: the kPa (kilopascal) is the official unit accepted in the International system of units to express pressure.

25.09 Gauge Pressure

This is the pressure above atmospheric pressure. When stating gauge pressure of a gas the plus or minus sign must also be shown to indicate pressure or vacuum.

25.10 Hydraulic Radius

$$\text{hydraulic radius} = \frac{\text{cross-sectional area}}{\text{wetted perimeter}}$$

25.11 Pressure Loss in Any Pipe

These formulas apply to any liquid.

$$h_1 = \frac{fLv^2}{D2g}$$

$$\Delta P = \frac{pfLv^2}{144D2g}$$

where

h_1 = loss of static head due to friction and flow (ft)
f = friction factor
L = length of pipe (ft)
v = mean velocity (ft/s)
D = diameter of pipe (internal diameter ft)
g = grav. constant = 32.2
p = specific weight (lb/ft^3)
ΔP = pressure drop (psi)

25.12 Friction Factor

$$f = \frac{64}{R_e}$$

$$f = \frac{64\mu_e}{Dv\text{p}}$$

$$f = \frac{64\mu}{124dvp}$$

f = friction factor
R_e = Reynolds number
μ_e = viscosity (lb/ft/s)
μ = absolute viscosity (centipoise)
D = diameter of pipe, internal diameter (ft)
v = mean velocity (ft/s)
p = density of fluid (lb/ft^3)

25.13 Poiseuille's Law for Laminar Flow

$$\Delta P = 0.000668 \frac{\mu L v}{d^2}$$

d = diameter of pipe (in.)
ΔP = pressure drop (psi)
L = length of pipe (ft)

25.14 Reynolds Number

The Reynolds number expresses the nature of the flow. When $R_e < 2100$ = laminar flow; when $R_e > 4000$ = turbulent flow.

$$R_e = \frac{\rho v D}{g\mu'_e}$$

$$R_e = 3160 \frac{Q}{v_k d}$$

μ'_e = absolute viscosity (slugs/ft/s)
v_k = kinematic viscosity (centi-stokes)

25.15 Critical Velocity

In fluid flow, the critical velocity is found at a Reynold's numbers of 2000–4000, i.e., when the flow changes from laminar to turbulent.

25.16 Total Head

$$H = z + \frac{144p}{\rho} + \frac{V^2}{2g}$$

$$= z + h + \frac{V^2}{2g}$$

$$= z + h + h_v$$

25.17 Pressure Head

$$h = \frac{144p}{\rho}$$

25.18 Velocity Head (Loss of Static Head)

$$h_v = \frac{V^2}{2g}$$

25.19 Resistance Coefficient

Resistance to flow due to valves, elbows, etc.

$$k = f\frac{L}{D}$$

$$k = \frac{h_l}{V^2/2g} = \frac{h_l 2g}{V^2}$$

H = total head (ft)
z = potential head above reference level, i.e., difference in elevation (ft)
p = pressure (psi gauge)
ρ = fluid density (lb/ft^3)
V = mean velocity (ft/min)
g = gravity constant (32.2)
h = pressure head (ft)
h_v = velocity head (ft)
f = friction factor
L = length of pipe
D = diameter of pipe
h_l = head loss (see **25.21**)
k = resistance coefficient

25.20 Bernoulli's Theorem

$$z_1 + h_1 + \frac{V_1^2}{2g} = z_2 + h_2 + \frac{V_2^2}{2g} + h_l$$

25.21 Head Loss

$$h_l = \frac{144 \Delta P}{\rho}$$

25.22 Flow Coefficient of Valves

$$C_v = \frac{29.9 d^2}{k^{1/2}}$$

25.23 Flow Through a Valve

Condition: viscosity similar to water.

$$Q = C_v \left[\Delta P \left(\frac{62.4}{\rho} \right) \right]^{1/2}$$

25.24 Pressure Drop Through Valves

$$\Delta P = \frac{\rho}{62.4} \left(\frac{Q}{C_v} \right)^2$$

z = elevation head
h = pressure head
h_l = pressure loss
V = mean velocity
C_v = flow coefficient for valves
d = diameter, (in.)
k = resistance coefficient of valve
Q = quantity (gal/min)
ΔP = pressure drop (psi)
ρ = density (lb/ft^3)

25.25 Flow Through Pipe

$$Q_1 = Av$$

Note: Q can also be expressed in ft^3/min but v must then also be expressed in ft/min.

Q_1 = quantity (ft^3/s)
A = area (ft^2)
v = mean velocity (ft/s)

25.26 Velocity vs. Cross-Sectional Area

$$\frac{v_1}{v_2} = \frac{A_2}{A_1}$$

25.27 Potential Energy for Fluids

$$E_e = W'z$$

$$E_p = W'h$$

$$E_v = \frac{W'v^2}{2g}$$

25.28 Total Energy of a Liquid

$$E_t = E_e + E_p + E_v$$

E_t = total energy
E_e = energy due to elevation (ft-lb)

E_p = energy due to pressure
E_v = kinetic energy (ft-lb)
W' = weight of mass (lb)

25.29 Power of a Liquid

Rate at which a liquid can do work.

$$P_e = Wz$$

$$P_p = Wh$$

$$P_v = \frac{WV^2}{2g}$$

$$P_t = P_e + P_p + P_v$$

$$hp = \frac{P_t}{550}$$

P_e = elevation power (ft-lb/s)
P_p = pressure power
P_v = velocity power
P_t = total power
W = weight of liquid per unit time (lb/s)

25.30 Flow Through Nozzles and Orifices

$$Q = CA(2gh_l)^{1/2}$$

$$Q = CA\,[(2g144\Delta P)/p]^{1/2}$$

25.31 Flow Coefficient

$$C = \frac{C_d}{\left[1 - \left(\frac{d_o}{d_1}\right)^4\right]^{1/2}}$$

Q = rate of flow (ft^3/s)
C = flow coefficient
A = area of orifice (ft^2)
h_l = loss of head due to flow (ft)
ΔP = pressure drop (psi)
p = density (lb/ft^3)
C_d = discharge coefficient
d_o = diameter of orifice
d_1 = diameter of pipe upstream

25.32 Flow Through Pipes

$$Q = 19.65d^2\left(\frac{h_l}{k}\right)^{1/2}$$

k = resistance coefficient
d = diameter of pipe, internal diameter (ft)
q = flow rate (ft^3/s)
L = length of crest
h_o = weir head
g = constant (32.2)

25.33 Flow Through Rectangular Weir

$$q = 0.415(L - 0.2h_o)h_o^{1.5}(2g)^{1/2}$$

If $h_o > L$, then

$$q = 0.386Lh_o^{1.5}(2g)^{1/2}$$

25.34 Flow Through Triangular Weir

$$q = \frac{0.31h_o^{2.5}(2g)^{1/2}}{\tan\phi}$$

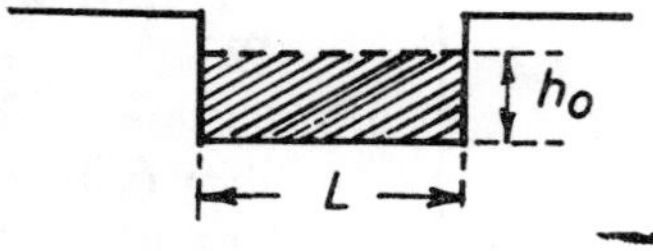

25.35 Gas Flow Measurements

a) Location of sampling ports

For accurate measurements, the sample ports should be located from one half to two duct diameters *upstream* and two to eight duct diameters *downstream* from disturbances such as bends, reductions, and others.

b) Minimum of traverse points

To obtain accurate results, a minimum of 12 traverse points are required in the duct cross section as shown in the following sketch:

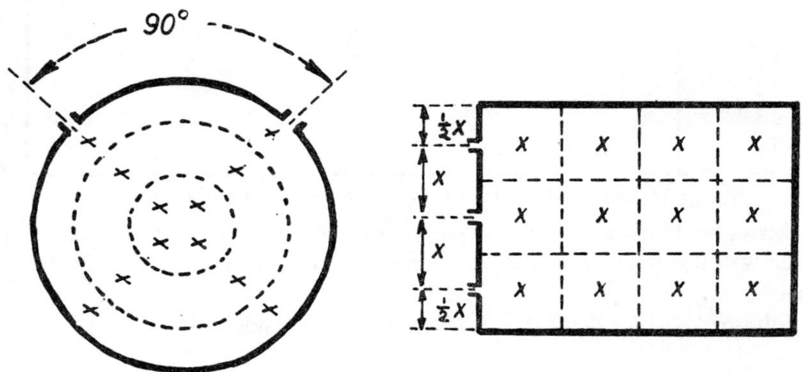

c) Traverse points for circular stacks

To cover equal areas in a circular stack or duct and thus obtain an accurate traverse, the following table can be used to locate individual points in the traverse.

Traverse point	*Total traverse points per diameter* 6	8	10	12	14
1	.044	.033	.025	.021	.018
2	.147	.105	.082	.067	.057
3	.295	.194	.146	.118	.099
4	.705	.323	.226	.177	.146
5	.853	.677	.342	.250	.201
6	.956	.806	.658	.355	.269
7		.895	.774	.645	.366
8		.967	.854	.750	.634
9			.918	.823	.731
10			.975	.882	.799
11				.933	.854
12				.979	.901
13					.943
14					.982

Numbers indicated are fractional distances of the diameter.

Example: When a 12-point traverse has to be made on a stack having an inside wall diameter of 72 inches, where must the fourth traverse point be located?

Solution: (.177)(72) = 12.7 in. from the inside wall (*ans.*)

25.36 Pitot Tube Measurements

$$V = 1096.7\left(\frac{\Delta p}{d}\right)^{1/2}$$

where

$$d = 1.325\left(\frac{\text{BAR}}{t + 460}\right)$$

$$Q = AV$$

Note: When the pressure of the gas is 29.92 in. Hg and the density equal to air then:

$$V = 174[\Delta p(t + 460)]^{1/2}$$

V = velocity (fpm)
Δp = differential pressure (in. H_2O)
d = gas density (lb/ft^3)
BAR = absolute pressure inside duct (in. Hg)
t = gas temperature (°F)
Q = flow rate (cfm)
A = duct cross section (ft^2)

25.37 S-Tube Measurements

The S-tube operates on the same principle as the pitot tube but is primarily used to measure gases that contain dust particles which would have the tendency to plug a pitot tube. To use the S-tube, the tube factor (c_s) has to be known or be determined by calibration. Normally, c_s = 0.85–0.90. The velocity is calculated by the following formula where "Δp" and "d" have the same meaning as in the preceding formula.

$$V = c_s 1096.7(\Delta p/d)^{1/2}$$

25.38 One-Point Traverse

In cases where only approximate flow rate determinations are required, a one-point reading of the differential pressure can be made in the center of a circular duct. However, the calculated velocity (V) has to be multiplied by a factor of 0.91 to obtain the approximate average velocity in the duct.

25.39 Conversions of Flow Rates

a) English system of units

$$Q_{scf} = Q_{acf}\left(\frac{520}{t + 460}\right)\left(\frac{\text{in. Hg}}{29.92}\right)$$

$$Q_{acf} = Q_{scf}\left(\frac{t + 460}{520}\right)\left(\frac{29.92}{\text{in. Hg}}\right)$$

b) Metric system of units

$$Nm^3 = m^3{}_{(act)}\left(\frac{273}{T + 273}\right)\left(\frac{\text{mm Hg}}{760}\right)$$

$$m^3{}_{(act)} = Nm^3 \left(\frac{T + 273}{273}\right)\left(\frac{760}{\text{mm Hg}}\right)$$

where

Q_{scf} = volume at standard condition (60°F, sea level)
Nm^3 = volume at standard condition (0°C, sea level)
t = temperature of gas (°F)
T = temperature of gas (°C)

25.40 Flow Determination with Orifice Plate

In small diameter pipes, an orifice plate is usually more conveniently employed than a pitot tube to measure flow rates.

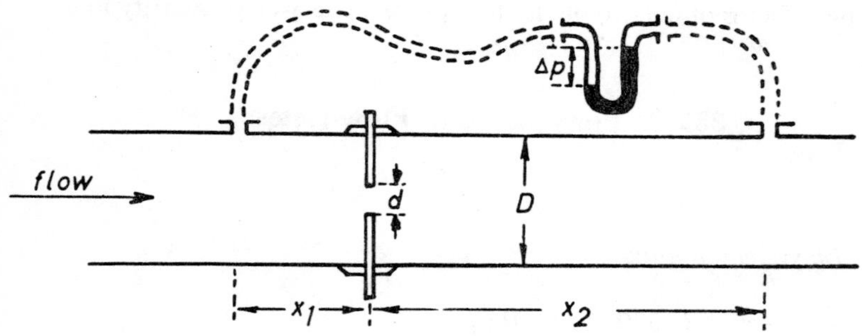

Rules to follow

$\frac{d}{D}$ = > 0.7
m = < 0.02D
x_1 = 2.5D
x_2 = 8.0D

For air only, the velocity is calculated by the formula

$$V = 174c_o[\Delta p(t + 460)]^{1/2}$$

For gases other than air, use the formula given in **25.30**. The flow coefficient (c_o) can be found in the following table.

Values for c_o

A_2/A_1	0.1	0.2	0.3	0.4	0.5	0.6	0.7
3 in. pipe	.619	.631	.653	.684	.728	.788	.880
6 in. pipe	.616	.627	.648	.677	.719	.777	.869
12 in. pipe	.610	.620	.637	.663	.700	.756	.848

V = velocity (fpm)
p = differential pressure (in. H_2O)
t = temperature of air (°F)
A_1 = area of pipe (ft^2)
A_2 = area of orifice (ft^2)

25.41 Ventury Meters

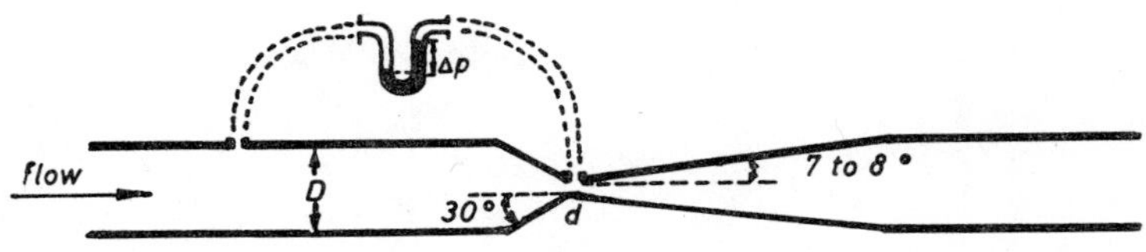

(Note: d = 0.5–0.33D)

For air only

$$V = 174c_v[\Delta p(t + 460)]^{1/2}$$

where

$$c_v = 0.98\left[\frac{1}{1 - (A_2/A_1)^2}\right]^{1/2}$$

V = velocity (fpm)
p = differential pressure (in. H_2O)
t = temperature (°F)
A_1 = area of pipe at "D" (ft^2)
A_2 = area of throat (ft^2)

Chapter 26

HEAT TRANSFER

Symbols used:

A = cross-sectional area (ft^2 or m^2)

A_1 = area at right angle to direction of heat flow (ft^2 or m^2)

a = radiation area (ft^2)

a_c = surface area in contact with the gas

C_p = specific heat (Btu/lb°F)

c = specific heat (kcal/kg°C or Btu/lb°F)

c_g = specific heat of gas (kcal/kg°C or Btu/lb°F)

c_s = specific heat of solid (kcal/kg°C or Btu/lb°F)

d = distance (ft or m)

D = inside diameter of pipe (ft)

$\frac{dQ}{d\theta}$ = quantity of heat transferred per unit time

$\frac{dt}{dx}$ = rate of temperature change with distance in the direction of the heat flow.

d_c = heat transfer coefficient

H_{12} = radiant interchange of heat of two bodies (Btu/h ft^2)

H = heat radiated (Btu/h ft^2)

h = fluid film coefficient (Btu/h ft^2 °F)
H_q = rate of heat flow (Btu/h)
I = intensity of radiation (cal/s cm^2)
k = coefficient of thermal conductivity
k_1 = natural convection coefficient
k_2 = thermal conductivity (Btu-ft/h ft^2 °F)
Q = quantity of heat (Btu or kcal)
Q_θ = quantity of heat transferred in unit time
Q_1 = quantity of heat received (English or metric units)
Q_r = quantity of heat radiated (English or metric units)
t_1 = initial temperature (F, C, R, or K)
t_2 = final temperature (F, C, R, or K)
T = absolute temperature (Kelvin)
T_a = absolute temperature (Rankine)
T_1 = absolute temperature of hotter body (Rankine)
T_2 = absolute temperature of colder body (Rankine)
Δt_1 = least temperature difference
Δt_2 = largest temperature difference
t_g = temperature of the gas
t_s = temperature of the solid
w = weight of body (English or metric units)
w_g = weight of gas (English or metric units)
w_s = weight of solid (English or metric units)
x = emissivity factor
X_a = area factor
y = Stefan-Boltzmann constant = 0.174×10^{-8}
μ = absolute viscosity (lb/ft h)
σ = constant = 1.36×10^{-12}
θ = time

26.01 Heat Required for a Temperature Change

This formula can be used in English or the metric system of units:

$$Q = cw(t_2 - t_1)$$

26.02 Conduction

This is defined as the transfer of heat within a substance or from one substance to another while they are in contact with each other. Use English or the metric system of units.

$$Q = \frac{kA}{d}(t_2 - t_1)$$

Fourier's law:

$$\frac{dQ}{d\theta} = -kA_1 \frac{dt}{dx}$$

26.03 Convection

This is defined as the transfer of heat by the motion of the particles of the heated substance itself. Convection occurs only in liquids and gases by circulation. This formula is applicable to the English and the metric system:

$$Q = k_1 A(t_2 - t_1)$$

26.04 Radiation

This is defined as the transfer of heat from one body to another without the two being in contact with each other.

$$Q_1 = \frac{Q_r}{d^2}$$

Stefan-Boltzmann law

This law expresses the relationship between the intensity of radiation and the absolute temperature of the body. For a "black" body

$$I = \sigma T^4$$
$$H_q = ya(T_a)^4$$
$$H_q = 0.174a(T_a/100)^4$$

Emissivity and absorptivity

$$H = 0.174 \times a(T_a/100)^4$$

Note: A perfect black body is the best emitter of radiant energy. In other words, $x = 1.0$. For all other substances, $x = > 0$ and < 1.

Heat exchange between two radiant bodies

$$H_{12} = 0.174X_t(T_1/100)^4\,(T_2/100)^4$$

Likewise, the heat radiated from a specific area:

$$H_{12} = 0.174X_tX_a(T_1/100)^4\,(T_2/100)^4$$

where

$$X_t = \frac{1}{\dfrac{1}{x_1} + \dfrac{1}{x_2} - 1}$$

Mean radiant temperature

This is defined as the mean temperature of all the surrounding surfaces.

$$MRT = \frac{A_1 t_1 + A_2 t_2 + \ldots\ldots + A_n t_n}{A_1 + A_2 + \ldots\ldots + A_n}$$

Logarithmic mean temperature difference

$$LMTD = 0.434 \frac{\Delta t_2 - \Delta t_1}{\log(\Delta t_2/\Delta t_1)}$$

26.05 Nusselt Number (N_u)

$$N_u = \frac{hD}{k_2}$$

26.06 Prandtl Number (P_r)

$$P_r = \frac{C_p \mu}{k_2}$$

26.07 General Heat Transfer Equations

Gas to a solid

$$Q_\theta = d_c(T_2 - T_1)a_c\theta$$

If gas is used to cool a solid

$$w_g c_g(t_{g_2} - t_{g_1}) = w_s c_s(t_{s_1} - t_{s_2})$$

If gas is used to heat a solid

$$w_g c_g(t_{g_1} - t_{g_2}) = w_s c_s(t_{s_2} - t_{s_1})$$

26.08 Temperature Equivalents

$$°F = \frac{9}{5}°C + 32$$

$$°C = \frac{5}{9}(°F - 32)$$

Absolute temperatures

Rankine = °F + 459.6

Kelvin = °C + 273

26.09 Temperature Conversion Table

The number in the center of each column can be converted to either Centigrade or Fahrenheit units by using the appropriate value given to the left or right of this number. Example: convert 40°F to Centigrade: on the left = 4.4°C. Likewise, convert 40°C to Fahrenheit: on the right = 104°F.

°C		°F	°C		°F
–45.6	–50	–58	–12.8	9	48.3
–40.0	–40	–40	–12.2	10	50.0
–34.4	–30	–22	–11.7	11	51.8
–28.9	–20	– 4	–11.1	12	53.6
–23.3	–10	14	–10.6	13	55.4
–17.8	0	32	–10.0	14	57.2
–17.2	1	33.8	– 9.4	15	59.0
–16.7	2	35.6	– 8.9	16	60.8
–16.1	3	37.4	– 8.3	17	62.6
–15.6	4	39.2	– 7.8	18	64.4
–15.0	5	41.0	– 7.2	19	66.2
–14.4	6	42.8	– 6.7	20	68.0
–13.9	7	44.6	– 6.1	21	69.8
–13.3	8	46.4	– 5.6	22	71.6

°C		°F	°C		°F
– 5.0	23	73.4	17.8	64	147.2
– 4.4	24	75.2	18.3	65	149.0
– 3.9	25	77.0	18.9	66	150.8
– 3.3	26	78.8	19.4	67	152.6
– 2.8	27	80.6	20.0	68	154.4
– 2.2	28	82.4	20.6	69	156.2
– 1.7	29	84.2	21.1	70	158.0
– 1.1	30	86.0	21.7	71	159.8
– 0.6	31	87.8	22.2	72	161.6
0	32	89.6	22.8	73	163.4
0.6	33	91.4	23.3	74	165.2
1.1	34	93.2	23.9	75	167.0
1.7	35	95.0	24.4	76	168.8
2.2	36	96.8	25.0	77	170.6
2.8	37	98.6	25.6	78	172.4
3.3	38	100.4	26.1	79	174.2
3.9	39	102.2	26.7	80	176.0
4.4	40	104.0	27.2	81	177.8
5.0	41	105.8	27.8	82	179.6
5.6	42	107.6	28.3	83	181.4
6.1	43	109.4	28.9	84	183.2
6.7	44	111.2	29.4	85	185.0
7.2	45	113.0	30.0	86	186.8
7.8	46	114.8	30.6	87	188.6
8.3	47	116.6	31.1	88	190.4
8.9	48	118.4	31.7	89	192.2
9.4	49	120.2	32.2	90	194.0
10.0	50	122.0	32.8	91	195.8
10.6	51	123.8	33.3	92	197.6
11.1	52	125.6	33.9	93	199.4
11.7	53	127.4	34.4	94	201.2
12.2	54	129.2	35.0	95	203.0
12.8	55	131.0	35.6	96	204.8
13.3	56	132.8	36.1	97	206.6
13.9	57	134.6	36.7	98	208.4
14.4	58	136.4	37.2	99	210.2
15.0	59	138.3	37.8	100	212
15.6	60	140.0	43.3	110	230
16.1	61	141.8	48.9	120	248
16.7	62	143.6	54.4	130	266
17.2	63	145.4	60.0	140	284

°C		°F	°C		°F
65.6	150	302	293	560	1040
71.1	160	320	299	570	1058
76.7	170	338	304	580	1076
82.2	180	356	310	590	1094
87.8	190	374	316	600	1112
93.3	200	392	321	610	1130
98.9	210	410	327	620	1148
104.4	220	428	332	630	1166
110.0	230	446	338	640	1184
115.6	240	464	343	650	1202
121.1	250	482	349	660	1220
127	260	500	355	670	1238
132	270	518	360	680	1256
138	280	536	366	690	1274
143	290	554	371	700	1292
149	300	572	377	710	1310
154	310	590	382	720	1328
160	320	608	388	730	1346
166	330	626	393	740	1364
171	340	644	399	750	1382
177	350	662	404	760	1400
182	360	680	410	770	1418
188	370	698	416	780	1436
193	380	716	421	790	1454
199	390	734	427	800	1472
204	400	752	454	850	1562
210	410	770	482	900	1652
216	420	788	510	950	1742
221	430	806	538	1000	1832
227	440	824	566	1050	1922
232	450	842	593	1100	2012
238	460	860	621	1150	2102
243	470	878	649	1200	2192
249	480	896	677	1250	2282
254	490	914	704	1300	2372
260	500	932	732	1350	2462
266	510	950	760	1400	2552
271	520	968	788	1450	2642
277	530	986	815	1500	2732
282	540	1004	843	1550	2822
288	550	1022	871	1600	2912

°C		°F	°C		°F
899	1650	3002	1288	2350	4262
927	1700	3092	1316	2400	4352
954	1750	3182	1343	2450	4442
982	1800	3272	1371	2500	4532
1010	1850	3362	1399	2550	4622
1038	1900	3452	1427	2600	4712
1066	1950	3542	1454	2650	4802
1094	2000	3632	1482	2700	4892
1121	2050	3722	1510	2750	4982
1149	2100	3812	1538	2800	5072
1177	2150	3902	1593	2900	5252
1204	2200	3992	1649	3000	5432
1232	2250	4082	1704	3100	5612
1260	2300	4172	1759	3200	5792

For interpolation

°C		°F
0.56	1	1.8

Chapter 27

PHYSICAL CHEMISTRY

GASES

27.01 Gas Laws

These well-known laws apply to the English as well as the metric system of units.

Boyle's law: temperature constant

$$P_1 V_1 = P_2 V_2$$

$$P_2 = \frac{P_1 V_1}{V_2}$$

$$V_2 = \frac{P_1 V_1}{P_2}$$

Charles law: pressure constant

$$T_1 V_2 = T_2 V_1$$

$$V_2 = \frac{T_2 V_1}{T_1}$$

$$T_2 = \frac{T_1 V_2}{V_1}$$

General gas law

$$P_2 V_2 T_1 = P_1 V_1 T_2$$

P = absolute pressure
V = volume
T = absolute temperature
(R = °F + 460)
(K = °C + 273)

27.02 Gas Law Constant

$$R = \frac{PV}{T} = \text{constant}$$

R = gas constant

R = 0.0821 liter-atm/°K

R = 0.73 ft^3–atm/°R lb-mole

R = 1544 ft-lb/°R lb-mole

For any gas

$$PV = \frac{w}{M}RT$$

P = absolute pressure
V = volume
T = absolute temperature
R = gas constant
w = weight of gas
M = molecular weight

27.03 Avogadros Law

A mole of any substance contains the same number of molecules. Equal volumes of all gases under the same temperature and pressure conditions contain the same number of molecules:

$$\frac{w_1}{w_2} = \frac{M_1}{M_2}$$

The number of molecules in a mole of any substance is constant. Avogadro's number = 6.02×10^{23} (at standard condition). Also: 22.4 liters of any gas at standard condition contains the above number of molecules.

w = weight (g)

M = molecular weight

27.04 Density of a Gas

$$d = \frac{PM}{(1000)(0.0821)T}$$

27.05 Standard Condition of a Gas

At 0°C, 1 atm.

27.06 Normal Density of a Gas

d_o = g/liter at 0°C, 1 atm.

27.07 Molecular Weight of Gases

$$M = 22.41d_o$$

$$= \frac{1000d(0.0821)T}{P}$$

$$= \frac{w(0.0821)T}{Pv}$$

27.08 Density Changes of Gases

$$d_2 = d_1 \frac{T_1}{T_2} \frac{P_2}{P_1}$$

$$d = \frac{PM}{RT}$$

d = density (g/cm^3)
d_o = normal density (g/liter)
P = absolute pressure (atm)
T = absolute temperature (K)

R = gas constant

v = volume (liter)

27.09 Moles

mole = w/M

mole fraction = moles/total moles

mole percent = (mole fraction) 100

27.10 Volume Changes of a Gas

$$V_2 = V_1\left(\frac{T_2}{T_1}\right)\left(\frac{P_1}{P_2}\right)$$

$$V_1 = V_2\left(\frac{T_1}{T_2}\right)\left(\frac{P_2}{P_1}\right)$$

SOLUTIONS

27.11 Weight Percent of Solutions

This is defined as the number of grams of solute per 100 grams of solvent.

Example: 14 grams of salt dissolved in 100 grams of water gives 114 grams of solution. Thus,

$$\text{solute} = \frac{14}{114}(100) = 12.28 \text{ percent}$$

In many instances, the weight of solute is expressed also in terms of the volume of solution, e.g., 14 grams of salt per liter of solution.

27.12 Mole Fraction of a Solution

This is defined as the number of gram molecules (moles) of solute per total number of moles contained in the solution.

Example: 14 grams of NaCl (sodium chloride) are dissolved in 100 grams of water. (Note: molecular weight of NaCl = 58.46 and water = 18.02). Thus,

$$\text{for NaCl,} \quad \frac{14}{58.46} = 0.239 \text{ moles}$$

$$\text{for water,} \quad \frac{100}{18.02} = 5.549 \text{ moles}$$

The total number of moles in solution = 0.239 + 5.549 = 5.788. Hence the mole fraction of the solute is

$$\frac{0.239}{5.788} = 0.0413 \quad (\textit{ans.})$$

27.13 Molality of a Solution

The number of moles of solute per liter of water.
Example: 140 grams of NaCl per 1000 grams of water.

$$\frac{140}{58.46} = 2.395 \text{ molal}$$

27.14 Molarity of a Solution

A molar solution contains one mole of solute per liter of solution.

Example: The molecular weight of NaCl (sodium chloride) is 58.46. Thus a molar table salt solution contains 58.46 grams of NaCl per liter of solution.

SOLIDS

27.15 Percent of an Element Contained in a Compound

$$\text{percent} = \frac{\text{atomic weight of element}}{\text{molecular weight of compound}}(100)$$

Example: What percent of iron is contained in Fe_2O_3?
Solution: atomic weight of Fe = 55.85
molecular weight of Fe_2O_3 = 159.7
Since there are 2 atoms of Fe in the compound:

$$\text{percent} = \frac{(2)(55.85)}{159.7}(100) = 69.94 \text{ percent} \quad (\textit{ans.})$$

27.16 Percent of a Compound Contained in a Substance

$$\text{percent} = \frac{\text{molecular weight of compound}}{\text{molecular weight of substance}}(100)$$

Example: What percent of calcium oxide are contained in calcium carbonate?
Solution: mol. wt. of CaO = 56.08
mol. wt. of $CaCO_3$ = 100.09, therefore

$$\text{percent} = \frac{56.08}{100.09}(100) = 56.029 \quad (\textit{ans.})$$

27.17 Weight Problems

Calculations involving weight and mixture problems can best be solved by writing the problem in the form of a chemical equation as shown in the following example.

Example: How much calcium oxide, CaO, and how much sulfur trioxide, SO_3, is required to produce 168 kg of calcium sulfate, $CaSO_4$?

Solution:

mol. wt →	56.08		80.06		136.14
	CaO	+	SO_3	→	$CaSO_4$
kg →	x_1		x_2		168.0

1. kg of CaO needed:

$$\frac{56.08}{x_1} = \frac{136.14}{168}, \qquad x_1 = 69.20 \quad (ans.)$$

2. kg of SO_3 needed:

$$\frac{80.06}{x_2} = \frac{136.14}{168}, \qquad x_2 = 98.80 \quad (ans.)$$

or

$$x_2 = 168 - 69.20 = 98.80$$

Chapter 28

PHYSICS

28.01 Newton's Law of Gravitation

$$F = g\,\frac{m_1 m_2}{d^2}$$

where

F	=	gravitational force
m_1, m_2	=	mass of bodies
d	=	distance between the centers of gravity of the two bodies.
g	=	gravitational constant = 32.2 ft/s^2
		= 981 cm/s^2

28.02 Acceleration—Forces

The absolute unit is the dyne which expresses the force that produces acceleration, i.e., a change in momentum on a body at rest or in motion.

Metric unit: dyne = the force required to produce an acceleration of 1 cm/s^2 in a gram mass.

English units: poundal = the force required to produce an acceleration of 1 ft/s^2 to a pound mass.

$$\frac{F_1}{a_1} = \frac{F_2}{a_2}$$

$$a = \frac{F}{m}$$

F = force
a = acceleration
m = mass

28.03 Mass of a Body

Mass expresses the quantity of matter. The metric unit for mass is the gram, the English unit is the pound.

$$m = \frac{F}{a}$$

28.04 Weight of a Body

Weight is defined as the force with which a body is attracted toward the earth.

$$w = mg$$

28.05 Work Done

$$W = Fs$$

Work is expressed as the product of the force acting on a body and the distance the body has moved against the resistance.

Metric system of units:

W = erg = force of one dyne through one centimeter
= joule = 1.0×10^7 ergs
= g/cm = 980 ergs

English system of units:

W = ft-lb = force required to move a mass of one pound a distance of one foot.
= ft-poundal = one poundal acting through one foot.

28.06 Power

The time rate at which work is done.

$$P = \frac{W}{t}$$

Metric units

$$\text{watts} = \frac{\text{joules}}{\text{seconds}}$$

English units

$$\text{horsepower} = \frac{\text{ft-lb/min}}{33{,}000}$$

$$= \frac{\text{ft-lb/s}}{550}$$

Conversions

one watt	=	1.0×10^7 ergs/s
one kilowatt	=	1000 watt
one hp	=	550 ft-lb/s
	=	33,000 ft-lb/min
	=	746 watt

28.07 Molecular Heat of Gases

This is defined as the heat required to raise the temperature of one gram-mole of a gas one degree Celsius.

$c_p = c_v + R$

$c_v = (3/2)R$ (for monatomic gases only).

where

c_p = molecular heat at constant pressure
c_v = molecular heat at constant temperature
R = gas constant = 1.988 cal/°K

28.08 Molecular Heat of Solids

$m = x_a s_a + x_b s_b + \ldots\ldots$

$m = Ms$

where

m = molecular heat of solid
M = molecular weight
s = specific heat
x = atomic weight

28.09 Latent Heat of Fusion

This is defined as the heat required to obtain a change of state without a temperature change in the substance. Latent heat of fusion is expressed in either cal/mole, cal/g, or Btu/lb.

Example

$$H_2O_{ice} \longrightarrow H_2O_{liq.} \quad = \ -144 \text{ Btu/lb}$$
$$= \ -80 \text{ cal/g}$$
$$= \ -1558 \text{ cal/mole}$$

28.10 Latent Heat of Evaporation

This is defined as the heat required to change a substance from a liquid to a gaseous state without a change in temperature. This is also known as the enthalpy of evaporation.

Example

$$H_2O_{liq.} \longrightarrow H_2O_{vapor} \quad = \ 970.2 \text{ Btu/lb}$$
$$= \ 539 \text{ cal/g}$$
$$= \ 10{,}500 \text{ cal/mole}$$

28.11 Heat of Formation and Reaction

This is defined as the heat units absorbed or evolved in a chemical reaction to form one mole of a substance.

$$AB + CD = AC + BD + h$$

When the heat of formation of individual compounds is known, e.g.,

$$A + B = AB + a, \text{etc.,}$$

then

$$AB + CD = AC + BD - a - b + c + d$$

and

$$h = c + d - (a + b)$$

where

A,B,C	=	compounds weight
a,b,c	=	heat of formation
h	=	heat of reaction

28.12 Joule Equivalent

This is the mechanical equivalent of heat.

$$J = \frac{\text{kcal}}{426.9}$$

$$J = \frac{\text{Btu}}{1694.1}$$

where

J = joule equivalent (kg/m)

28.13 Temperature of a Mixture

$$T_{final} = \frac{w_1 c_1 t_1 + w_2 c_2 t_2 + \ldots}{w_1 c_1 + w_2 c_2 + \ldots}$$

28.14 Gas Mixtures

$$p_f = p_1 + p_2 + p_3 + \ldots$$

$$w_f = w_1 + w_2 + w_3 + \ldots$$

$$c_v = w_1 c_{v_1} + w_2 c_{v_2} + \ldots$$

$$c_p = w_1 c_{p_1} + w_2 c_{p_2} + \ldots$$

$$V_f = V_1 + V_2 + V_3 + \ldots$$

$$T_{af} = \frac{p_1 V_1 + p_2 V_2 + \ldots}{\frac{p_1 V_1}{T_{a_1}} + \frac{p_2 V_2}{T_{a_2}} + \ldots}$$

$$= \frac{p_1 V_1 + p_2 V_2 + \ldots}{RW}$$

28.15 Gas Constant, R

For air = 29.3
For O_2 = 26.5

c = specific heat
c_p = specific heat at constant pressure
c_v = specific heat at constant volume
P,p = pressure
w = weight
V,v = volume
T_a = absolute temperature
T_{af} = final absolute temperature
t = temperature
t_f = final temperature
R = gas constant (m-kg/f/kg/°C)

28.16 Friction Coefficient

This is defined as the ratio of the force required to move one body over the other to the total force pressing the two bodies together.

$$k = \frac{F}{Ft}$$

28.17 Moment of Force—Torque

This is the force that produces rotation about an axis.

$$L = Fd$$

where

L = torque (dyne-cm),
F = force that produces rotation about center (dyne), and
d = perpendicular distance from the line of action of the force to the axis (cm).

Chapter 29

PSYCHROMETRY

29.01 Basic Psychrometric Equation

$$p = p' - AP(t - t')$$

where

$A = 6.60 \times 10^{-4}\ (1 + 0.00115t')$ when °C and mm Hg are used.
$A = 3.67 \times 10^{-4}\ [1 + 0.00064(t' - 32)]$ when °F and in. Hg are used.

29.02 Wet Bulb Depression

The wet bulb depression is expressed in either Celsius or Fahrenheit units.

$$\text{wet bulb depression} = t - t'$$

29.03 Relative Humidity

$$\text{relative humidity} = \frac{100p}{p_x}$$

p = partial pressure of water vapor at dry bulb temperature
p' = saturation pressure of water vapor at wet bulb temperature
P = total barometric pressure
t = dry bulb temperature
t' = wet bulb temperature
p_x = saturation pressure of water vapor at dry bulb temperature

29.04 Dew Point

When the partial pressure of water vapor at a stated temperature equals the saturation pressure of water vapor at the same temperature, the air is saturated, i.e., the dew point has been reached.

$$h_s = h + \frac{q}{970.2}(T - T_s)$$

h_s = saturation humidity, dew point
h = absolute humidity
q = $0.24 + 0.45h$ = heat capacity
T = dry bulb temperature (°F)
T_s = saturation temperature (°F)

29.05 Properties of Air and Water Vapor

English units
at 14.7 psia

Temperature (°F)	*Dry air* (lb/ft^3)	*Water vapor* (lb/ft^3)	*Vapor pressure* (psi)
30	.0811	.0504	.082
32	.0807	.0502	.089
34	.0804	.0500	.096
36	.0801	.0498	.104
38	.0797	.0496	.113
40	.0794	.0494	.122
42	.0791	.0492	.131
44	.0788	.0490	.142
46	.0785	.0488	.153
48	.0781	.0486	.165
50	.0779	.0484	.178
52	.0776	.0482	.192
54	.0773	.0481	.206
56	.0770	.0479	.222
58	.0767	.0477	.238
60	.0764	.0475	.256
62	.0761	.0473	.275
64	.0758	.0472	.295

English units (*cont'd*)

Temperature (°F)	*Dry air* (lb/ft^3)	*Water vapor* (lb/ft^3)	*Vapor pressure* (psi)
66	.0755	.0470	.316
68	.0752	.0468	.339
70	.0749	.0466	.363
72	.0746	.0465	.388
74	.0743	.0463	.416
76	.0741	.0461	.444
78	.0738	.0459	.475
80	.0735	.0458	.507
82	.0732	.0456	.541
84	.0730	.0454	.577
88	.0724	.0451	.656
90	.0722	.0450	.698
92	.0719	.0448	.743
94	.0717	.0446	.791
96	.0714	.0445	.841
98	.0711	.0443	.894
100	.0709	.0443	.949
102	.0706	.0440	1.008
104	.0704	.0439	1.071
106	.0701	.0437	1.136
108	.0699	.0436	1.204
110	.0696	.0434	1.276
112	.0694	.0433	1.351
114	.0692	.0431	1.431
116	.0689	.0430	1.513

Metric units

at 1 atmosphere

Temperature (°C)	*Dry air* (kg/m^3)	*Water vapor* (kg/m^3)	*Vapor pressure* (atm.)
0	1.293	.804	.0061
1.1	1.288	.801	.0065
2.2	1.283	.798	.0071
3.3	1.277	.795	.0077
4.4	1.272	.791	.0083

(*cont'd*)

Temperature (°C)	*Dry air* (kg/m³)	*Water vapor* (kg/m³)	*Vapor pressure* (atm.)
5.6	1.267	.788	.0089
6.7	1.262	.785	.0097
7.8	1.258	.782	.0104
8.9	1.251	.779	.0112
10.0	1.248	.775	.0121
11.1	1.243	.772	.0131
12.2	1.239	.771	.0140
13.3	1.234	.767	.0151
14.4	1.229	.764	.0162
15.6	1.224	.761	.0174
16.7	1.219	.758	.0187
17.8	1.214	.756	.0201
18.9	1.210	.753	.0215
20.0	1.205	.750	.0231
21.1	1.200	.747	.0247
22.2	1.195	.745	.0267
23.3	1.190	.742	.0283
24.4	1.187	.739	.0302
25.6	1.182	.735	.0323
26.7	1.177	.734	.0345
27.8	1.173	.731	.0368
28.9	1.169	.727	.0393
30.0	1.165	.726	.0419
31.1	1.160	.723	.0446
32.2	1.157	.721	.0475
33.3	1.152	.718	.0506
34.4	1.149	.714	.0538
35.6	1.144	.713	.0572
36.7	1.139	.710	.0608
37.8	1.136	.708	.0646
38.9	1.131	.705	.0686
40.0	1.128	.703	.0729
41.1	1.123	.700	.0772
42.2	1.120	.698	.0819
43.3	1.115	.695	.0868
44.4	1.112	.694	.0919
45.5	1.109	.690	.0973
46.7	1.104	.689	.1030

PART V

EMISSION CONTROL AND PLANT EQUIPMENT

Chapter 30

TEST FOR PARTICULATE EMISSIONS

Formulas used to determine the particulate emission rate are given. They apply to tests performed with a dry gas meter. For details of the testing procedures, the reader is advised to refer to "Standard Performance for Stationary Sources," Federal Register, Dec. 23, 1971.

30.01 Data Needed for Stack Testing

A = area of stack at sample point (ft^2)

D = total weight of particulate collected (g)

d = nozzle diameter of sample tube (in.)

F_s = Pitot tube correction factor

CO_2 = percent CO_2 in stack gas (by volume, dry)

O_2 = percent O_2 in stack gas (by volume, dry)

N_2 = percent N_2 in stack gas (by volume, dry)

P_b = absolute barometric pressure (in. Hg)

P_o = average pressure drop across gas meter (in. H_2O)

P_s = absolute static stack pressure (in. Hg)

$\Delta P_{avg}{}^{1/2}$ = average square root of Pitot tube reading (in. H_2O)

T_s = absolute stack temperature (°R)

t = sampling time (min)

V_m = net meter volume recorded (ft^3)

w = total water collected (ml)

30.02 Summary of Results

G_d	=	average density of the stack gas, relative to air	.....
Grains/ACF	=	dust concentration in stack gas	.....
MV_m	=	wet meter volume at meter condition (ft^3)	.....
MV_s	=	wet meter volume at stack condition (ft^3)	.....
MV_{std}	=	dry meter volume at standard condition (ft^3 @ 70°F, 1 atm.)	.....
M_w	=	average molecular weight of stack gas	.....
M_f	=	percent moisture in stack gas (decimal)	.....
P_m	=	average absolute meter pressure (in. Hg)	.....
Q_s	=	actual stack gas flow rate (ACFM)	.....
V_s	=	average stack gas velocity (fps)	.....
V_v	=	water vapor volume at meter condition (ft)	.....

30.03 Calculations

a) Conversion of water collected to gas at meter conditions
(for gas meter, internally corrected to 70°F)

$$P_m = P_b - \left(\frac{P_o}{13.6}\right) = \text{.....} \text{ in. Hg}$$

$$V_v = 1.415\left(\frac{w}{P_m}\right) = \text{.....} \text{ ft}^3$$

b) Percent moisture in flue gas

$$M_f = \frac{V_v}{V_v + V_m} = \text{.....} \text{ percent (expressed as a decimal)}$$

c) Density of gas relative to air

Step 1 Determination of molecular weight of gas:

$$CO_2 = 44(1 - M_f)\,\frac{CO_2}{100} = \ldots\ldots$$

$$O_2 = 32(1 - M_f)\,\frac{O_2}{100} = \ldots\ldots$$

$$N_2 = 28(1 - M_f)\,\frac{N_2}{100} = \ldots\ldots$$

$$H_2O = 18M_f = \ldots\ldots$$

$$M_w = \text{Total: } \ldots\ldots$$

Step 2 Determination of density relative to air:

$$G_d = \frac{M_w}{28.95} = \ldots\ldots$$

d) Average velocity of gas

$$V_s = 2.9F_s\left(\frac{29.92\,T_s}{P_s G_d}\right)^{1/2} \cdot (\Delta p_{avg})^{1/2} = \ldots\ldots \text{ft/s}$$

e) Stack gas flow rate

$$Q_s = V_s A\,60 = \ldots\ldots \text{ ACFM}$$

f) Conversion of wet meter volume to stack conditions

$$MV_m = V_v + V_m = \ldots\ldots \text{ ft}^3$$

$$MV_s = MV_m\left(\frac{P_m}{P_s}\right)\left(\frac{T_s}{530}\right) = \ldots\ldots \text{ ft}^3$$

g) Conversion of dry meter volume to standard conditions

$$MV_{\text{std}} = V_m\left(\frac{P_m}{29.92}\right) = \;.\,.\,.\,.\,.\;\text{DSCF}$$

h) Grain loading

$$\text{Grains/ACF} = 15.43\left(\frac{D}{MV_s}\right) = \;.\,.\,.\,.\,.$$

$$\text{Grains/DSCF} = 15.43\left(\frac{D}{MV_{\text{std}}}\right) = \;.\,.\,.\,.\,.$$

i) Emission rate

$$\text{lb/h} = DQ_s\left(\frac{60}{MV_s}\right)454 = \;.\,.\,.\,.\,.$$

j) Percent isokinetic sampling

$$\text{percent iso} = \frac{305.6MV_s}{V_s t d^2} = \;.\,.\,.\,.\,.\;\text{percent}$$

Chapter 31

USEFUL DATA FOR EMISSION CONTROL

31.01 Molecular Weights of Selected Gases

Gas	*Formula*	*Molecular weight*
Ammonia	NH_3	17.03
Carbon dioxide	CO_2	44.01
Carbon disulfide	CS_2	76.14
Carbon monoxide	CO	28.01
Chlorine	Cl_2	70.91
Fluorine	F_2	38.00
Hydrogen chloride	HCl	36.47
Hydrogen	H_2	2.016
Hydrogen fluoride	HF	20.01
Hydrogen sulfide	H_2S	34.08
Nitric oxide	NO	30.01
Nitrogen	N_2	28.02
Nitrogen dioxide	NO_2	46.01
Nitrous oxide	N_2O	44.02
Oxygen	O_2	32.00
Sulfur dioxide	SO_2	64.07

31.02 Conversion Factors for Emission Rates

Multiply	*by*	*to obtain*
g/s	3.6	kg/h
g/s	7.9367	lb/h
g/s	190.48	lb/day
g/s	34.763	short tons/yr
g/s	86.4	kg/day
g/s	31.536	metric tons/yr
kg/h	0.27778	g/s
kg/h	8.76	metric tons/yr
kg/h	52.911	lb/day
lb/h	0.126	g/s
lb/h	0.45359	kg/h
lb/h	4.38	short tons/yr

Chapter 32

STORAGE AND TRANSPORT EQUIPMENT

32.01 Drum Dryers

a) Evaporation rate

$$R = \frac{WM}{1 - M} 2000$$

where

$$M = m_i - m_o$$

b) Dryer volume

$$V = 0.7854D^2 L$$

c) Specific rate of evaporation

$$s_v = \frac{R}{V}$$

R = evaporation rate (lb H_2O/h)
W = feed rate (tph)
m_i = moisture of feed at dryer inlet (decimal)
m_o = moisture of feed at dryer outlet (decimal)
D = internal diameter of dryer (ft)
L = internal length of dryer (ft)
V = internal volume of dryer (ft^3)
s_v = specific evaporation rate (lb H_2O/h/ft^3)

d) Specific heat consumption

$$s_h = \frac{Q}{R}$$

e) Feed retention time (for approximation only)

$$t = \frac{20.56L}{SND}$$

s_h = specific heat consumption (Btu/lb water evaporation)
Q = firing rate (Btu/h)
t = retention time (min)
S = slope of dryer (degrees)
N = speed of dryer (rpm)

32.02 Slurry Pumps

a) Specific gravity of slurry

The specific gravity of the slurry can be obtained directly from the table given in **3.01** or, when the specific gravity of the dry solids is not 2.70, it can be calculated by the following formula:

$$s_g = \frac{100s_d}{(100 - M_f) + M_f s_d}$$

b) Slurry pumping rate

In **3.04** a formula is given that uses the pulp density of the slurry as a variable. Another formula that is useful:

$$Q = \frac{W_c W_a 0.2}{(100 - M_f)s_g}$$

s_g = specific gravity of slurry
s_d = specific gravity of dry solids
W_c = clinker output (tph)
W_a = lb dry solids per ton of clinker
M_f = slurry moisture content (percent)
Q = slurry flow rate (gpm)

c) Power required for pumping

$$hp = \frac{Qh_t s_g}{3960 p_e m_e}$$

hp = horsepower required to pump slurry
h_t = total head (ft)
p_e = pump efficiency
m_e = mechanical and electrical efficiency of motor and drive

d) Friction factor for pipe lines

$$F_f = \frac{0.9y}{s_g d}$$

Note: Use 0.85 if y is not known.

F_f = friction factor (ft/ft)
y = yield stress (lb/ft^2)
d = internal pipe diameter (in.)

32.03 Bucket Elevators

a) Elevator capacity

$$C = 1.8 \frac{Vv_e fd}{s}$$

Note: f = normally .45 to .65

C = capacity (tph)
V = bucket volume (ft^3)
v_e = elevator velocity (fps)
s = bucket spacing (ft)
f = bucket load factor
d = bulk density of material (lb/ft^3)

b) Elevator, horsepower required

$$hp = 1.1 \frac{C(33 + h \sin \theta)}{760w}$$

w = bucket width (ft)
h = height of elevator (ft, sprocket to sprocket)
θ = angle of inclination

32.04 Belt Conveyors

a) Conveyor capacity

$$C = Avdn \frac{60}{2000}$$

Note: $A = 0.09r^2$

b) Horsepower required for belt conveyor

$$hp = 0.08148(C)^{1/3}\,[5.7 + (L)^{1/2}] \pm \ldots\ldots$$

$$\ldots \pm 0.001242C \sin \theta\,(33 + L)$$

For horizontal conveyors:

$$hp = 0.57967(C)^{1/3}$$

A = cross-sectional area of material on belt (ft^2)
v = belt speed (fpm)
r = material width on belt (ft)
n = efficiency (.8–.9)
L = conveyor belt length (ft, pulley to pulley)

C = capacity (tph)
d = density of material (lb/ft^3)
θ = angle of inclination

32.05 Screw Conveyors

a) Screw conveyor capacity

$$C = 0.005625ADnkd$$

k, performance factor	
conveyor length (ft)	k
10	.995
20	.98
30	.97
40	.96
50	.945
60	.93
70	.92
80	.91
90	.895
100	.88

b) Horsepower required

$$hp = (L + 30)\left(\frac{C}{582} \pm \frac{C \sin\theta}{270}\right)1.1$$

A = cross-sectional area of helix (ft^2)
D = diameter of helix (ft)
n = conveyor speed (rpm)
k = performance factor
d = bulk density of material (lb/ft^3)
C = capacity (tph)
L = conveyor length (ft)
θ = conveyor angle of inclination

32.06 Water Pumps

a) Centrifugal pump head

$$H = \left(\frac{D_1 n}{1900}\right)^2$$

H = head developed by pump (ft of H_2O)
D_1 = diameter of impeller (in.)
n = impeller speed (rpm)

b) Horsepower required to pump water

$$hp = \frac{QHs}{3960e}$$

Q = water flow rate (gpm)
hp = horsepower needed to drive pump
s = specific gravity of water
e = pump efficiency (decimal)

32.07 Storage Tanks

a) Contents in vertical cylindrical tanks

$$Q = 7.481D^2H_l$$

Note: This formula applies only to flat bottom tanks.

C = tank content (gal)
D = tank diameter (ft)
H_l = height of liquid (ft)

b) Contents in cones

$$V = 0.3333R^2 h$$

$$C = 2.49367R^2 h$$

V = volume of cone (ft^3)
R = radius of cone (ft)
h = height of cone (ft)

c) Compressed air receivers

The minimum size receiver to be used is

$$v_{\min} = \frac{dp_i}{p_o + 14.7}$$

$v_{\min}$ = minimum volume required (ft^3)
d = first-stage displacement of compressor (cfm)
p_i = inlet pressure (psi absolute)
p_o = outlet pressure (psi absolute)

32.08 Drag Chains

a) Drag chain capacity

$$C = 1.8whskd$$

Performance factor, k	
drag chain length (ft)	*k*
20	.995
40	.978
60	.963
80	.946
100	.930
120	.914
140	.898
160	.882
180	.866
200	.850

b) Horsepower required

$$hp = \left(\frac{L + 16.4}{3.3}\right)\left[\frac{(g + 0.8kw_m)n}{132}\right] \pm \frac{C \sin\theta}{270}$$

C = capacity (tph)
w = width of conveyor (ft)
h = depth of material (ft)
s = drag chain speed (ft/s)
k = performance factor
d = bulk density of material conveyed (lb/ft^3)
L = length of conveyor
g = weight of chain (lb/ft)
w_m = weight of material in drag chain (lb/ft)
n = drag chain speed (ft/s)
θ = angle of inclination

32.09 Jaw and Gyratory Crushers

a) Jaw crusher capacity

$$Q = fdlwjna0.8$$

FACTOR "*f*"

Corrugated plates	
Type of rock	*f*
normal	3.05×10^{-5}
screened	2.54×10^{-5}
large	2.07×10^{-5}

Smooth plates	
Type of rock	*f*
normal	4.14×10^{-5}
screened	3.64×10^{-5}
large	3.12×10^{-5}

Correction factor "a"					
Jaw angle (°)	26	24	22	20	18
"*a*" =	1.0	1.06	1.12	1.18	1.24

b) Horsepower required

According to Viard's formula

$$hp = 0.1w_2 s$$

Q = capacity (tph)
f = factor (see table)
d = density (lb/ft^3)
l = length of discharge opening (in.)
w = width of discharge opening (in.)
j = length of jaw amplitude (in.)
n = strokes per min
a = correction factor for jaw angle
hp = motor size required
w_2 = width of swing jaw (in.)
s = maximum feed size (in.)

32.10 Stacks and Chimneys

a) Theoretical draft

$$d_t = 0.256hP_B \frac{1}{T_a + 460} - \frac{1}{T_s + 460}$$

b) Draft loss

$$d_l = 0.0942\left(\frac{T_s + 460}{D^4}\right)\left(\frac{W}{100{,}000}\right)^2\left(1 + \frac{0.022h}{D}\right)$$

c) Available (natural) draft

$$d_{\text{nat.}} = d_t - d_l$$

where

d_t = theoretical draft (in. H_2O)
d_l = draft loss (in. H_2O)
h = stack height (ft)
P_B = barometric pressure (in. Hg)
T_a = ambient air temperature (°F)
T_s = average stack temperature (°F)
D = diameter of stack (ft)
W = gas flow rate (lb/h)

PART VI

APPENDIX

Section A

MATHEMATICS

ALGEBRA

A1.01 Exponents

$$(a^m)(a^n) = a^{m+n}$$

$$(a^m)^n = a^{mn}$$

$$(a^m)(b^m) = (ab)^m$$

$$\frac{a^m}{a^n} = a^{m-n}$$

$$\frac{a^n}{b^n} = \left(\frac{a}{b}\right)^n$$

$$a^{1/2} = \sqrt{a}$$

$$a^{1/3} = \sqrt[3]{a}$$

$$a^{1/4} = \sqrt[4]{a}$$

$$a^k = \sqrt[k]{a}$$

$$a^{3/2} = \sqrt{a^3}$$

$$a^{-n} = \frac{1}{a^n}$$

A1.02 Fractions

$$\frac{a}{b} \pm \frac{c}{b} = \frac{a \pm c}{b}$$

$$\frac{a}{b} \times \frac{c}{d} = \frac{ac}{bd}$$

$$\frac{a}{b} \div \frac{c}{d} = \frac{ad}{bc} = \frac{a}{b} \times \frac{d}{c}$$

A1.03 Radicals

$$\left(\sqrt[n]{a}\right)^n = a$$

$$\sqrt[n]{a^n} = a$$

$$\sqrt[n]{a}\ \sqrt[n]{b} = \sqrt[n]{ab}$$

$$\frac{\sqrt[n]{a}}{\sqrt[n]{b}} = \sqrt[n]{\frac{a}{b}}$$

A1.04 Factoring

$$ax + ay = a(x + y)$$

$$a^2 - b^2 = (a + b)(a - b)$$

$$a^2 + 2ab + b^2 = (a + b)^2$$

$$a^2 - 2ab + b^2 = (a - b)^2$$

$$a^3 + b^3 = (a + b)(a^2 - ab + b^2)$$

$$a^3 - b^3 = (a - b)(a^2 + ab + b^2)$$

A1.05 Scientific Notations

1.2×10^2	=	120
1.2×10^3	=	1200
1.2×10^4	=	12000
1.2×10^{-1}	=	0.12
1.2×10^{-2}	=	0.012
1.2×10^{-3}	=	0.0012

A1.06 Logarithms

$$\log x + \log y = \log (xy)$$

$$\log x - \log y = \log \frac{x}{y}$$

$$x \log y = \log (y^x)$$

$$\log a^n = n \log a$$

$$\log \sqrt[n]{x} = \frac{1}{n} \log x$$

$$\log_a b = \frac{1}{\log_b a}$$

$$\log_{10} N = 0.4343 \log_e N$$

$$\log_e N = 2.3026 \log_{10} N$$

$$\text{In } 3.16 = \log_e 3.16$$

$$\log_b N^n = n \log_b N$$

A1.07 Determinants

Simultaneous equations:

$$ax + by + cz = d$$
$$ex + fy + gz = h$$
$$ix + jy + kz = l$$

Solutions

$$x = \frac{dfk + bgl + cjh - cfl - gjd - khb}{afk + bgi + cje - cfi - gja - keb}$$

$$y = \frac{ahk + dgi + cle - chi - gla - ked}{afk + bgi + cje - cfi - gja - keb}$$

$$z = \frac{afl + bhi + dje - dfi - hja - leb}{afk + bgi + cje - cfi - gja - keb}$$

A1.08 Quadratic Equation

$$ax^2 + bx + c = 0$$

$$x = \frac{-b \pm \sqrt{b^2 - 4ac}}{2}$$

If

$b^2 - 4ac$ is > 0, the roots are real and unequal.
$b^2 - 4ac$ is $= 0$, the roots are real and equal.
$b^2 - 4ac$ is < 0, the roots are imaginary.

A1.09 Powers of Ten

pico	= 10^{-12}		deka	= 10^{1}
nano	= 10^{-9}		hecto	= 10^{2}
micro	= 10^{-6}		kilo	= 10^{3}
milli	= 10^{-3}		mega	= 10^{6}
centi	= 10^{-2}		giga	= 10^{9}
deci	= 10^{-1}		tera	= 10^{12}

A1.10 Power and Roots

x	x^2	x^3	$\sqrt{x}$	$\sqrt[3]{x}$
1	1	1	1.000	1.000
2	4	8	1.414	1.260
3	9	27	1.732	1.442
4	16	64	2.000	1.587
5	25	125	2.236	1.710
6	36	216	2.449	1.817
7	49	343	2.646	1.913
8	64	512	2.828	2.000
9	81	729	3.000	2.080
10	100	1,000	3.162	2.154
11	121	1,331	3.317	2.224
12	144	1,728	3.464	2.289
13	169	2,197	3.606	2.351
14	196	2,744	3.742	2.410
15	225	3,375	3.873	2.466
16	256	4,096	4.000	2.520
17	289	4,913	4.123	2.571
18	324	5,832	4.243	2.621
19	361	6,859	4.359	2.668
20	400	8,000	4.472	2.714
21	441	9,261	4.583	2.759
22	484	10,648	4.690	2.802
23	529	12,167	4.796	2.844
24	576	13,824	4.899	2.884
25	625	15,625	5.000	2.924

(cont'd)

x	x^2	x^3	$\sqrt{x}$	$\sqrt[3]{x}$
26	676	17,576	5.099	2.962
27	729	19,683	5.196	3.000
28	784	21,952	5.291	3.037
29	841	24,389	5.385	3.072
30	900	27,000	5.477	3.107
31	961	29,791	5.568	3.141
32	1,024	32,768	5.657	3.175
33	1,089	35,937	5.745	3.208
34	1,156	39,304	5.831	3.240
35	1,225	42,875	5.916	3.271
36	1,296	46,656	6.000	3.302
37	1,369	50,653	6.083	3.332
38	1,444	54,872	6.164	3.362
39	1,521	59,319	6.245	3.391
40	1,600	64,000	6.325	3.420
41	1,681	68,921	6.403	3.448
42	1,764	74,088	6.481	3.476
43	1,849	79,507	6.557	3.503
44	1,936	85,184	6.633	3.530
45	2,025	91,125	6.708	3.557
46	2,116	97,336	6.782	3.583
47	2,209	103,823	6.856	3.609
48	2,304	110,592	6.928	3.634
49	2,401	117,649	7.000	3.659
50	2,500	125,000	7.071	3.684
51	2,601	132,651	7.141	3.708
52	2,704	140,608	7.211	3.733
53	2,809	148.877	7.280	3.756
54	2,916	157,464	7.348	3.780
55	3,025	166,375	7.416	3.803
56	3,136	175,616	7.483	3.826
57	3,249	185,193	7.550	3.849
58	3,364	195,112	7.616	3.871
59	3,481	205,379	7.681	3.893
60	3,600	216,000	7.746	3.915
61	3,721	226,981	7.810	3.936
62	3,844	238,328	7.874	3.958
63	3,969	250,047	7.937	3.979

(cont'd)

x	x^2	x^3	$\sqrt{x}$	$\sqrt[3]{x}$
64	4,096	262,144	8.000	4.000
65	4,225	274,625	8.062	4.021
66	4,356	287,496	8.124	4.041
67	4,489	300,763	8.185	4.062
68	4,624	314,432	8.246	4.082
69	4,761	328,509	8.307	4.102
70	4,900	343,000	8.367	4.121
71	5,041	357,911	8.426	4.141
72	5,184	373,248	8.485	4.160
73	5,329	389,017	8.544	4.179
74	5,476	405,224	8.602	4.198
75	5,625	421,875	8.660	4.217
76	5,776	438,976	8.718	4.236
77	5,929	456,533	8.775	4.254
78	6,084	474,552	8.832	4.273
79	6,241	493,039	8.888	4.291
80	6,400	512,000	8.944	4.309
81	6,561	531,441	9.000	4.327
82	6,724	551,368	9.055	4.344
83	6,889	571,787	9.110	4.362
84	7,056	592,704	9.165	4.380
85	7,225	614,125	9.220	4.397
86	7,396	636,056	9.274	4.414
87	7,569	658,503	9.327	4.431
88	7,744	681,472	9.381	4.448
89	7,921	704,969	9.434	4.465
90	8,100	729,000	9.487	4.481
91	8,281	753,571	9.539	4.498
92	8,464	778,688	9.592	4.514
93	8,649	804,357	9.644	4.531
94	8,836	830,584	9.695	4.547
95	9,025	857,375	9.747	4.563
96	9,216	884,736	9.798	4.579
97	9,409	912,673	9.849	4.595
98	9,604	941,192	9.899	4.610
99	9,801	970,299	9.950	4.626
100	10,000	1,000,000	10.000	4.642

A1.11 Fractions and Decimal Equivalents

1/32	=	0.03125	17/32	=	0.53125
1/16	=	0.0625	9/16	=	0.5625
3/32	=	0.09375	19/32	=	0.59375
1/8	=	0.125	5/8	=	0.625
5/32	=	0.15625	21/32	=	0.65625
3/16	=	0.1875	11/16	=	0.6875
7/32	=	0.21875	23/32	=	0.71875
¼	=	0.25	¾	=	0.75
9/32	=	0.28125	25/32	=	0.78125
5/16	=	0.3125	13/16	=	0.8125
11/32	=	0.34375	27/32	=	0.84375
3/8	=	0.375	7/8	=	0.875
13/32	=	0.40625	29/32	=	0.90625
7/16	=	0.4375	15/16	=	0.9375
15/32	=	0.46875	31/32	=	0.96875
½	=	0.5			

ft		in.		ft decimal	ft		in.		ft decimal
1/12	=	1	=	0.0833	7/12	=	7	=	0.5833
1/6	=	2	=	0.1667	2/3	=	8	=	0.6667
¼	=	3	=	0.25	¾	=	9	=	0.75
1/3	=	4	=	0.3333	5/6	=	10	=	0.8333
5/12	=	5	=	0.4167	11/12	=	11	=	0.9167
½	=	6	=	0.5					

TRIGONOMETRY

A2.01 Right Triangle

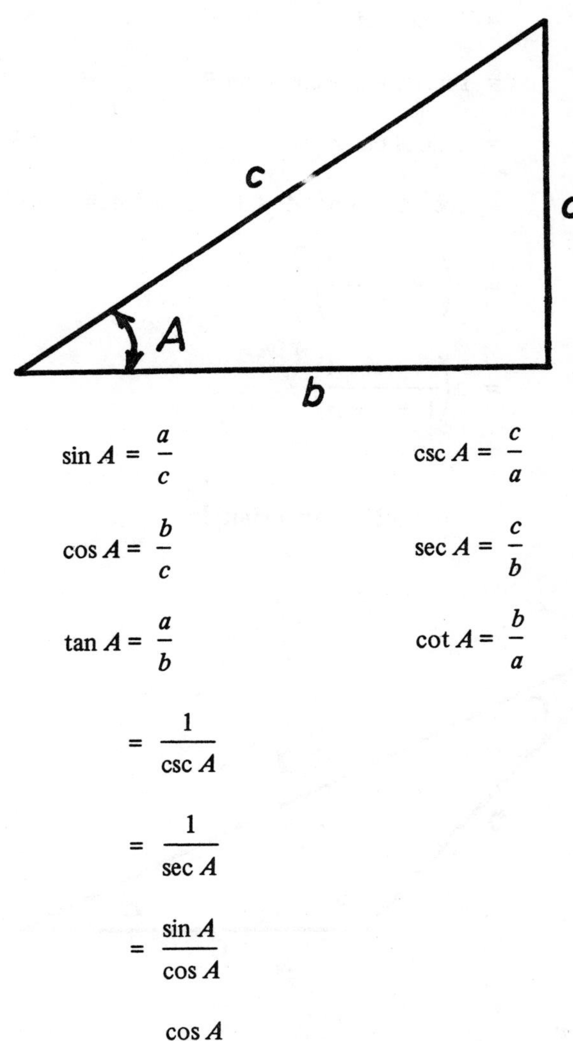

$$\sin A = \frac{a}{c} \qquad \csc A = \frac{c}{a}$$

$$\cos A = \frac{b}{c} \qquad \sec A = \frac{c}{b}$$

$$\tan A = \frac{a}{b} \qquad \cot A = \frac{b}{a}$$

$$\sin A = \frac{1}{\csc A}$$

$$\cos A = \frac{1}{\sec A}$$

$$\tan A = \frac{\sin A}{\cos A}$$

$$\cot A = \frac{\cos A}{\sin A}$$

$$\sin^2 A + \cos^2 A = 1$$

$$\sec^2 A = 1 + \tan^2 A$$

$$\csc^2 A = 1 + \cot^2 A$$

$$\sin (A + B) = \sin A \cos B + \cos A \sin B$$

$$\sin 2A = 2 \sin A \cos A$$

$$\cos 2A = \cos^2 A - \sin^2 A = 1 - 2 \sin^2 A = 2 \cos^2 A - 1$$

$$\sin \tfrac{1}{2}A = \pm\left(\frac{1 - \cos A}{2}\right)^{1/2}$$

$$\tan \tfrac{1}{2}A = \pm\left(\frac{1 - \cos A}{1 + \cos A}\right)^{1/2}$$

A2.02 Any Triangle

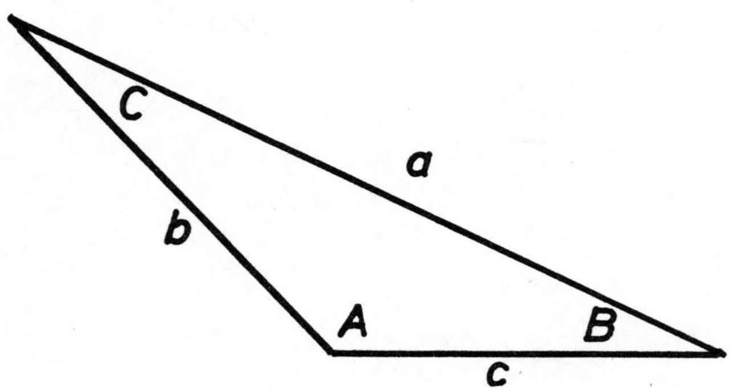

$$s = \frac{a + b + c}{2}$$

Law of sines

$$\frac{a}{\sin A} = \frac{b}{\sin B} = \frac{c}{\sin C}$$

Law of Cosines

$$a^2 = b^2 + c^2 - 2bc \cos A$$
$$b^2 = a^2 + c^2 - 2ac \cos B$$
$$c^2 = a^2 + b^2 - 2ab \cos C$$

Law of Tangents

$$\frac{\tan \frac{1}{2}(A-B)}{\tan \frac{1}{2}(A+B)} = \frac{a-b}{a+b} \qquad \text{where } a > b$$

$$\frac{\tan \frac{1}{2}(B-C)}{\tan \frac{1}{2}(B+C)} = \frac{b-c}{b+c} \qquad \text{where } b > c$$

$$\frac{\tan \frac{1}{2}(A-C)}{\tan \frac{1}{2}(A+C)} = \frac{a-c}{a+c} \qquad \text{where } a > c$$

Newton's Formula

$$\frac{c}{\sin \frac{1}{2}C} = \frac{a+b}{\cos \frac{1}{2}(A-B)}$$

Tangents of Half Angles

$$\tan \tfrac{1}{2}A = \frac{\left[\dfrac{(s-a)(s-b)(s-c)}{s}\right]^{1/2}}{(s-a)}$$

$$\tan \tfrac{1}{2}B = \frac{\left[\dfrac{(s-a)(s-b)(s-c)}{s}\right]^{1/2}}{(s-b)}$$

$$\tan \tfrac{1}{2}C = \frac{\left[\dfrac{(s-a)(s-b)(s-c)}{s}\right]^{1/2}}{(s-c)}$$

SOLUTIONS OF TRIANGLES

Given	*Use*
One side and two angles	Law of sines
Two sides and an angle opposite one of them	Law of sines
Two sides and their included angle	Law of tangents
Three sides	Tangent of half angles

Area

$$A = [s(s-a)(s-b)(s-c)]^{1/2}$$

$$A = \frac{a^2 \sin B \sin C}{2 \sin A}$$

$$A = \frac{b^2 \sin A \sin C}{2 \sin B}$$

$$A = \frac{c^2 \sin A \sin B}{2 \sin C}$$

$$A = \tfrac{1}{2} bc \sin A$$

$$A = \tfrac{1}{2} ac \sin B$$

$$A = \tfrac{1}{2} ab \sin C$$

STATISTICS

A3.01 Standard Deviation

When calculating the standard deviation of a population larger than 30 use the formula

$$s = \left[\frac{(x_1 - \bar{x})^2 + (x_2 - \bar{x})^2 + \ldots + (x_n - \bar{x})^2}{n}\right]^{1/2}$$

For samples whose total number is less than 30, ($n = < 30$) use

$$s = \left[\frac{(x_1 - \bar{x})^2 + (x_2 - \bar{x})^2 + \ldots + (x_n - \bar{x})^2}{n - 1}\right]^{1/2}$$

If "n" is sufficiently large and their distribution normal then

68.26 percent = ± s from mean
95 percent = ± $2s$ from mean
99.6 percent = ± $3s$ from mean

A3.02 Variance

$$\sigma^2 = \frac{(x_1 - \bar{x})^2 + (x_2 - \bar{x})^2 + \ldots + (x_n - \bar{x})^2}{n}$$

A3.03 Coefficient of Variation

$$c_v = 100 \frac{s}{\bar{x}}$$

A3.04 Relative Frequency

$$p = \frac{\text{number of samples in range}}{\text{number of samples out of range}}$$

A3.05 Geometric Mean

$$G_m = (x_1 x_2 \ldots x_n)^{\frac{1}{n}}$$

$$\log G_m = \frac{\log x_1 + \log x_2 + \ldots + \log x_n}{n}$$

A3.06 Least Squares

To determine the correlation between two variables solve the simultaneous equations for a and b

$$\Sigma x = n + \Sigma x^2 b$$

$$\Sigma xy = \Sigma xa + \Sigma x^2 b$$

where x and y are the values of the plotted variables and n = the number of samples.

After these values have been found, the best line to fit the plotted points (least square equation) becomes

$$Y = a + bx$$

A3.07 Coefficient of Correlation

The coefficient of correlation is a measure of the proximity of the plotted points on a graph to the straight line represented by the least square equation

$$r = \left[\frac{(a\Sigma y + b\Sigma xy) - \bar{Y}\Sigma y}{\Sigma y^2 - y\Sigma y} \right]^{1/2}$$

where

$$\bar{Y} = \frac{\Sigma y}{n}$$

When "r" approaches unity, there is a good correlation, when it approaches zero, there exists too wide a scatter to obtain a correlation.

Symbols used (in **A3.01** to **A3.05**)

- s = standard deviation
- $\bar{x}$ = mean
- x_1, x_2 = individual observations
- n = total number of observations
- σ^2 = variance of the samples
- p = relative frequency

FINANCES

A4.01 Compound Interest

$$S_n = Fe^{rn}$$

where

- S_n = value after "n" years due to compounding interest rate "r"
- F = initial amount invested

Example: What is the value after 21 years when \$1250 is invested at a compound interest rate of 5¼ percent?

$$S_n = 1250e^{(0.0525)(21)} = 3764.61 \quad (ans.)$$

A4.02 Total Annual Cash Flow

$$C = I + D$$
$$C = R - E$$
$$C = P + D$$

where

C = annual cash flow
I = annual income
D = annual depreciation
R = annual revenues
E = annual expenses
P = after tax profit

A4.03 After Tax Profit

$$P = (1.00 - t)(R - E)$$

where

P, R, and E have the same meaning as in **4.02** and
t = tax rate (expressed as a decimal)

Example: A company had \$1.25 million in revenues and \$0.95 million of expenses in one year. What is the after tax profit when the tax rate assessed is 50 percent?

$$P = (1.00 - 0.50)(1.25 - 0.95) \times 10^6 = 150{,}000 \quad (ans.)$$

A4.04 Straight Line Depreciation

$$D = \frac{I - L}{n}$$

where

D = straight line depreciation
I = initial investment value
L = expected salvage value at end of useful life
n = expected useful life

Example: An office copier is being purchased for a cost of \$5800. Its useful life is expected to be six years; after these six years its salvage value is estimated at \$1350. What is the annual depreciation on this copier when the straight line depreciation method is used?

$$D = \frac{5800 - 1350}{6} = 741.67 \quad (ans.)$$

A4.05 Double-Declining Balance Depreciation

$$D = \frac{2(I - d_c)}{n}$$

where

D = double-declining balance depreciation
I = initial investment value
d_c = cumulative depreciation charged in previous years
n = useful life

A4.06 Sum-of-Years Digit Depreciation

$$D = \frac{2(n - y + 1)}{n(n + 1)} \quad ;$$

where

D = sum-of-years digit depreciation
n = useful life
y = consecutive number of years from start of investment to year where D applies.

A4.07 Sixth-Tenth Factor

This factor is used to estimate the costs to replace an old by a new identical unit.

$$C_f = \left(\frac{a}{b}\right)^{0.6} C_o$$

where

C_f = new costs
C_o = old costs
a = new capacity
b = old capacity

A4.08 Value of an Investment After Depreciation

$$Q = pe^{-rn}$$

$$Q = p\left(1 - \frac{r}{k}\right)^{kn}$$

where

Q = final value
p = initial value when new
r = rate of annual depreciation (decimal)
n = number of years
k = number of times per year depreciation is figured.

A4.09 Return on Investment, *ROI*

$$ROI = \frac{\text{profits per year}}{\text{investment}}$$

$$ROI = \frac{\text{total savings} \div \text{years of life}}{\text{original investment}}$$

A4.10 Simple Compound Interest

$$S_{ci} = p(1 + i)^n$$

A4.11 Present Worth

Worth after *n* years

$$p_v = P(1 + i)^n$$

$$P = p_v \frac{1}{(1 + i)^n}$$

A4.12 Equal Payment Series Compound Amount

$$\text{sum}_n = R \frac{(1 + i)^n - 1}{i}$$

where

sum_n = worth after *n* years

R = annual payment

$$\text{present or equal value} = R\left[\frac{(1 + i)^n - 1}{i(1 + i)^n}\right]$$

Note: p_v = present worth

P,p = principal

i = interest

n = number of years

A4.13 Compound Interest Factors $(1 + i)^n$

n	Percent compound interest, i						
	5.00	5.25	5.50	5.75	6.00	6.25	6.50
1	1.050	1.0525	1.0550	1.0575	1.0600	1.0625	1.0659
2	1.1025	1.1078	1.1130	1.1183	1.1236	1.1289	1.1342
3	1.1576	1.1659	1.1742	1.1826	1.1910	1.1995	1.2079
4	1.2155	1.2271	1.2388	1.2506	1.2625	1.2744	1.2865
5	1.2763	1.2915	1.3070	1.3225	1.3382	1.3541	1.3701
6	1.3401	1.3594	1.3788	1.3986	1.4185	1.4387	1.4591
7	1.4071	1.4302	1.4547	1.4790	1.5036	1.5286	1.5540
8	1.4775	1.5058	1.5347	1.5640	1.5938	1.6242	1.6550
9	1.5513	1.5849	1.6191	1.6540	1.6895	1.7257	1.7626
10	1.6289	1.6681	1.7081	1.7491	1.7908	1.8335	1.8771
11	1.7103	1.7557	1.8021	1.8496	1.8983	1.9481	1.9992
12	1.7959	1.8478	1.9012	1.9560	2.0122	2.0699	2.1291
13	1.8856	1.9449	2.0058	2.0684	2.1329	2.1993	2.2675
14	1.9799	2.0470	2.1161	2.1874	2.2609	2.3367	2.4149
15	2.0789	2.1544	2.2325	2.3132	2.3966	2.4828	2.5718
16	2.1829	2.2675	2.3553	2.4462	2.5404	2.6379	2.7390
17	2.2920	2.3866	2.4848	2.5868	2.6928	2.8029	2.9170
18	2.4066	2.5119	2.6215	2.7356	2.8543	2.9780	3.1067
19	2.5270	2.6437	2.7656	2.8929	3.0256	3.1641	3.3086
20	2.6533	2.7825	2.9178	3.0592	3.2071	3.3619	3.5236
21	2.7860	2.9286	3.0782	3.2351	3.3996	3.5720	3.7527
22	2.9253	3.0824	3.2475	3.4211	3.6035	3.7952	3.9966
23	3.0715	3.2442	3.4262	3.6178	3.8197	4.0324	4.2564
24	3.2251	3.4145	3.6146	3.8259	4.0489	4.2844	4.5331
25	3.3864	3.5938	3.8134	4.0458	4.2919	4.5522	4.8277

cont'd

n	6.75	7.00	7.25	7.50	7.75	8.00	8.25
1	1.0675	1.0700	1.0725	1.0750	1.0775	1.0800	1.0825
2	1.1396	1.1449	1.1503	1.1556	1.1610	1.1664	1.1718
3	1.2165	1.2250	1.2336	1.2423	1.2510	1.2597	1.2685
4	1.2986	1.3108	1.3231	1.3355	1.3479	1.3605	1.3731
5	1.3862	1.4026	1.4190	1.4356	1.4524	1.4693	1.4864
6	1.4798	1.5007	1.5219	1.5433	1.5650	1.5869	1.6090
7	1.5797	1.6058	1.6322	1.6590	1.6862	1.7138	1.7418
8	1.6863	1.7182	1.7506	1.7835	1.8169	1.8509	1.8855
9	1.8002	1.8385	1.8775	1.9172	1.9577	1.9990	2.0410
10	1.9217	1.9672	2.0136	2.0610	2.1095	2.1589	2.2094
11	2.0514	2.1049	2.1596	2.2156	2.2730	2.3316	2.3917
12	2.1899	2.2522	2.3162	2.3818	2.4491	2.5182	2.5890
13	2.3377	2.4098	2.4841	2.5604	2.6389	2.7196	2.8026
14	2.4955	2.5786	2.6642	2.7524	2.8434	2.9372	3.0338
15	2.6639	2.7590	2.8573	2.9589	3.0638	3.1722	3.2841
16	2.8437	2.9522	3.0645	3.1808	3.3012	3.4259	3.5551
17	3.0357	3.1588	3.2867	3.4194	3.5571	3.7000	3.8483
18	3.2406	3.3799	3.5249	3.6758	3.8328	3.9960	4.1658
19	3.4593	3.6165	3.7805	3.9515	4.1298	4.3157	4.5095
20	3.6928	2.8697	4.0546	4.2479	4.4499	4.6610	4.8816
21	3.9421	4.1406	4.3485	4.5664	4.7947	5.0338	5.2843
22	4.2082	4.4304	4.6638	4.9089	5.1663	5.4365	5.7202
23	4.4922	4.7405	5.0019	5.2771	5.5667	5.8715	6.1922
24	4.7954	5.0724	5.3646	5.6729	5.9981	6.3412	6.7030
25	5.1191	5.4274	5.7535	6.0983	6.4630	6.8485	7.2560

A4.14 Discounted Cash Flow Factors

Year	10 percent	*DCF* 25 percent	40 percent
0	1.000	1.000	1.000
1	.952	.885	.824
2	.861	.689	.553
3	.779	.537	.37
4	.705	.418	.248
5	.638	.326	.166
6	.577	.254	.112
7	.522	.197	.075
8	.473	.154	.050
9	.428	.119	.034
10	.387	.092	.023
11	.350	.073	.015
12	.317	.057	.010
13	.287	.044	.007
14	.259	.034	.005
15	.235	.027	.003
16	.212	.021	.002
17	.192	.016	.001
18	.174	.013	.001
19	.157	.010	.001
20	.142	.008	—

A4.15 Deposit Calculation

This type of calculation is used to determine the value of an account after "n" years when an annual deposit of "z" is made to the account.

$$S_n = (I)(1 + i)^n + z\left[\frac{(1 + i)^n - 1}{i}\right]$$

where

S_n = value of account after "n" years
I = initial capital invested in account
i = interest rate (expressed as a decimal)
n = number of years
z = annual deposit into account

Example: What is the account balance after 15 years when the initial deposit is $1500.00, an annual amount of $240.00 is deposited, and the account pays 5½ percent compound interest?

$$S_n = (1500)(1 + 0.055)^{15} + 240\left[\frac{(1 + 0.055)^{15} - 1}{0.055}\right]$$

$$S_n = 8726.79 \quad (ans.)$$

SAFETY FORMULAS

A5.01 Accident Frequency Rate

$$f = \frac{n(1.0 \times 10^6)}{h}$$

Accident frequency rate is defined in terms of number of accidents per million man-hours worked.

f = frequency rate
n = number of accidents during period under investigation
h = number of man-hours worked during the same period

A5.02 Severity Rate

Accident severity rate is defined in terms of the number of days lost due to accidents per 1000 man-hours worked.

$$s = \frac{1000d}{h}$$

where

s = severity rate (days lost/1000 man-hours)
d = days lost in period
h = total man–hours worked in same period

A5.03 Safety Performance

An individual group, department, or plant safety performance can be stated in terms of another's group known standard performance as follows:

$$\text{percent frequency} = \frac{100f}{f_{\text{std.}}}$$

$$\text{percent severity} = \frac{100s}{s_{\text{std.}}}$$

where $f_{\text{std.}}$ and $s_{\text{std.}}$ are the frequency and severity rates of other groups performing similar duties.

PLANE AND SOLID GEOMETRY

Plane Figures

A6.01 Rectangle

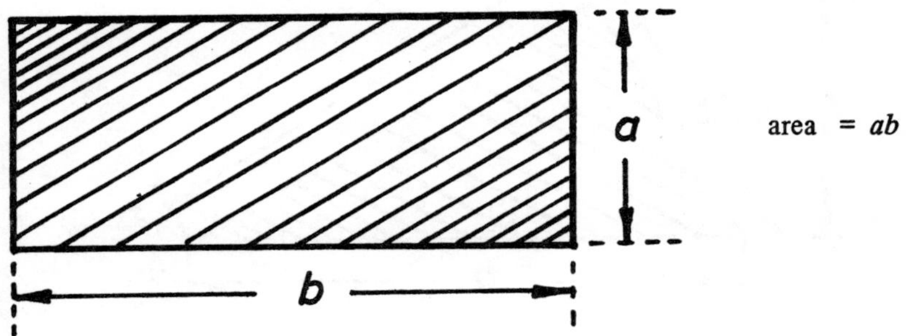

area = ab

A6.02 Parallelogram

area = ab

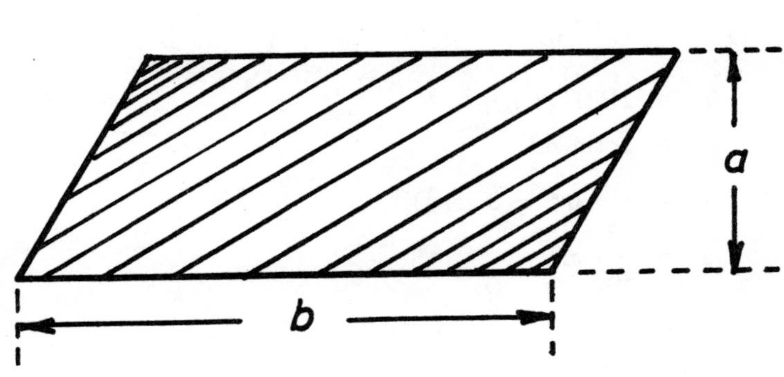

A6.03 Triangle

$$\text{area} = 0.5(ab)$$

Let $x = 0.5(b + c + d)$
then

$$\text{area} = [x(x - b)(x - c)(x - d)]^{1/2}$$

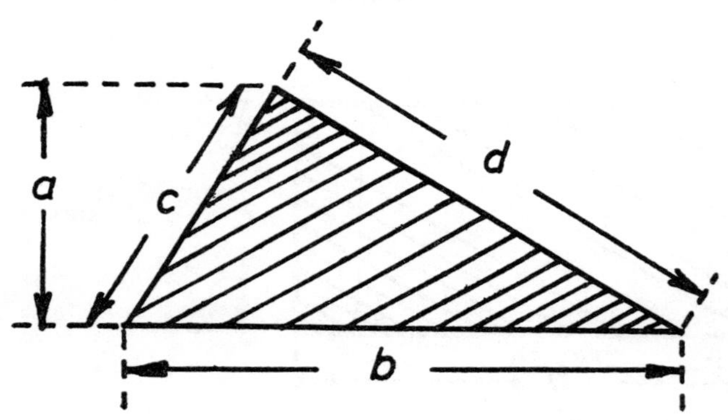

A6.04 Circle

$$\text{circumference} = \pi D = 2\pi r$$

$$\text{area} = 0.25\pi D^2 = \pi r^2$$

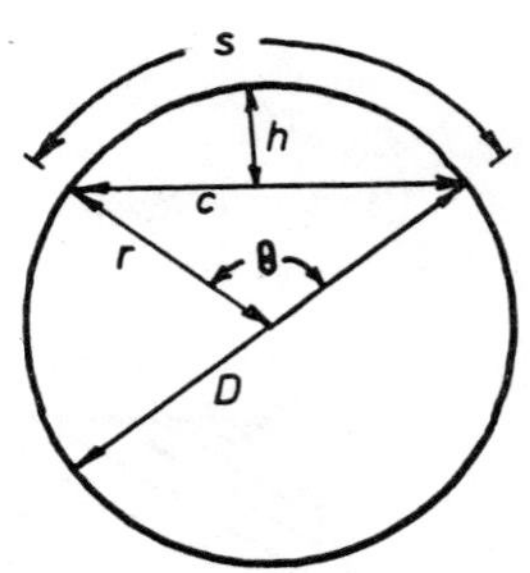

$$r = \frac{0.25c^2 + h^2}{2h}$$

$$c = 2[h(D - h)]^{1/2} = 2r \sin \tfrac{1}{2}\theta$$

$$h = r - (r^2 - 0.25c^2)^{1/2}$$

$$s = \frac{\theta}{360}\pi D = 0.01745r\theta$$

A6.05 Circular Sector

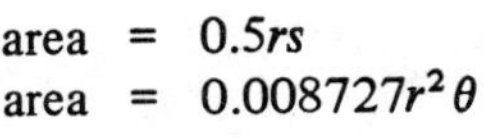

$$\text{area} = 0.5rs$$
$$\text{area} = 0.008727r^2\theta$$

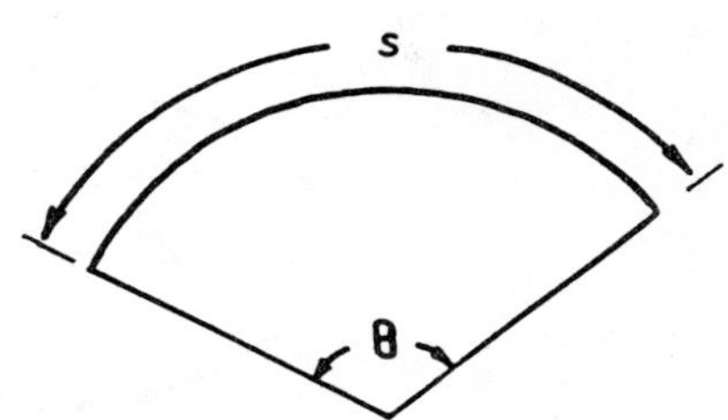

A6.06 Circular Segment

$$\text{area} = 0.5[rs - c(r - h)]$$

$$\text{area} = \pi r^2 \left(\frac{\theta}{360}\right) - \left(\frac{c(r-h)}{2}\right)$$

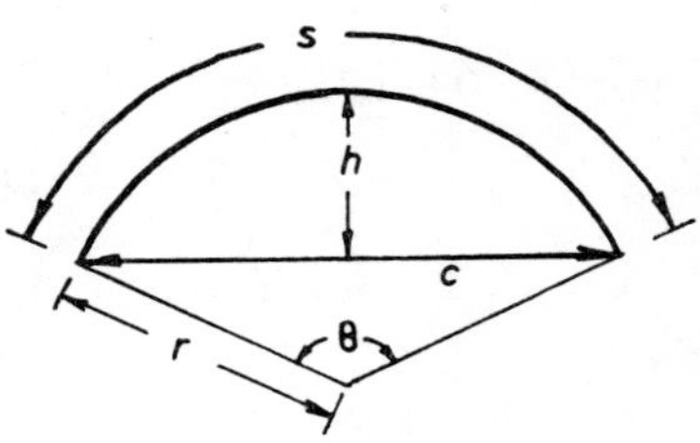

A6.07 Circular Ring

$$\text{area} = 0.7854(D^2 - d^2)$$

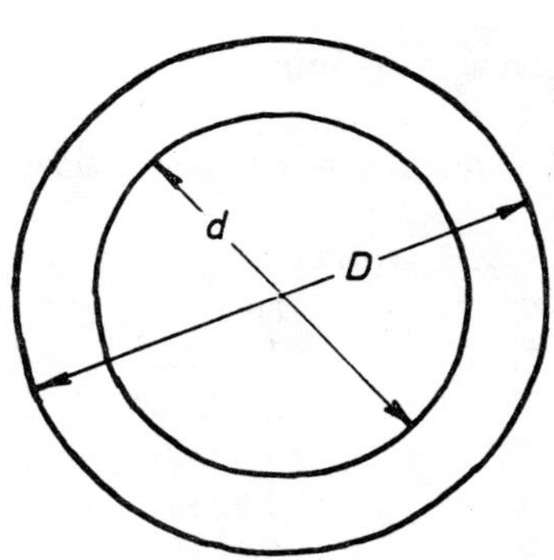

A6.08 Ellipse

area = $0.25\pi Aa$

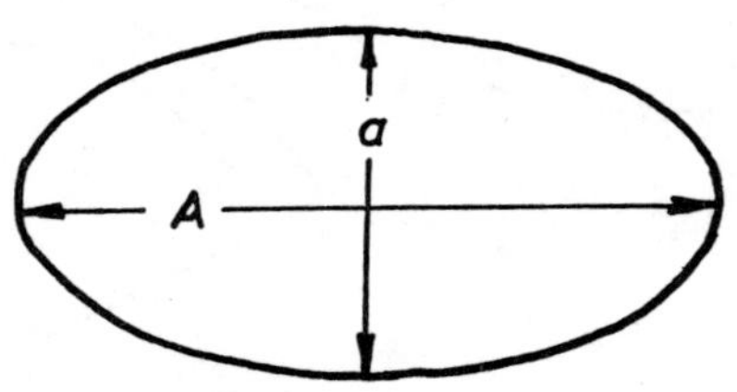

A6.09 Parabola

area = $0.6667ha$

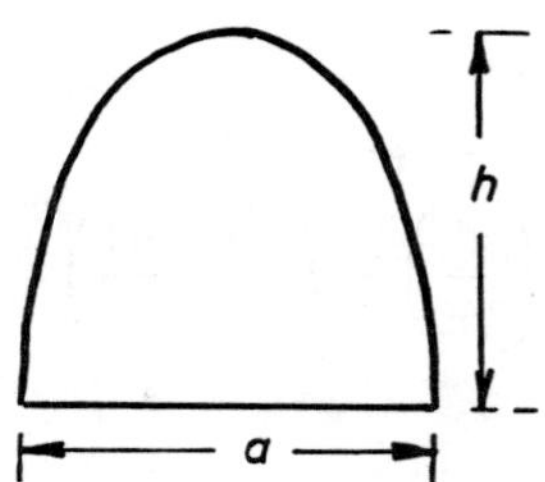

A6.10 Polygon

area = $0.5nsr$

where n = number of sides

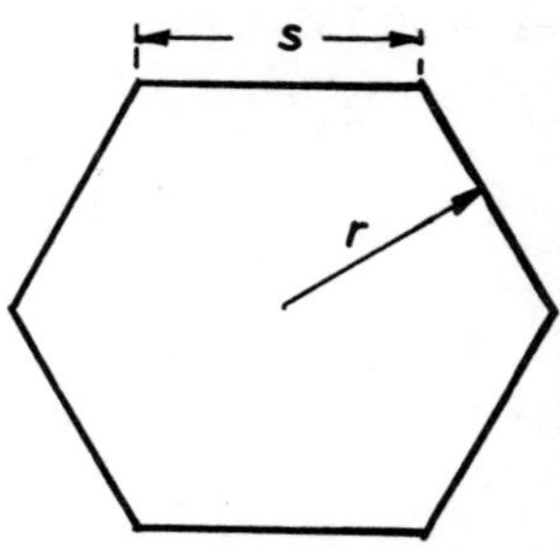

areas

n	area
5	$1.7205s^2$
6	$2.5981s^2$
7	$3.6339s^2$
8	$4.8284s^2$
9	$6.1818s^2$

A6.11 Trapezoid

area $= 0.5[b(H + h) + ch + aH]$

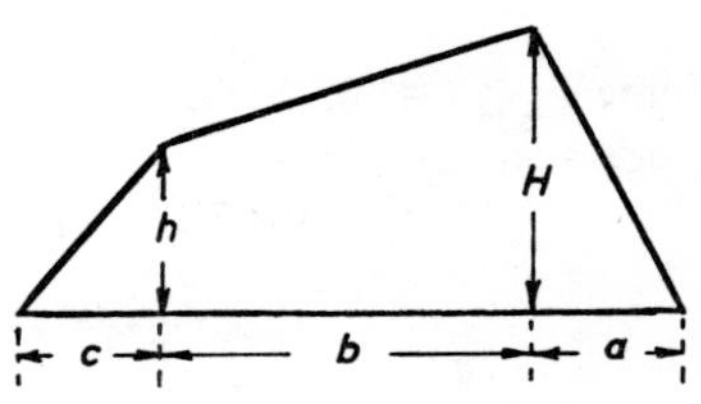

A6.12 Catenary

$y = a \cosh x$

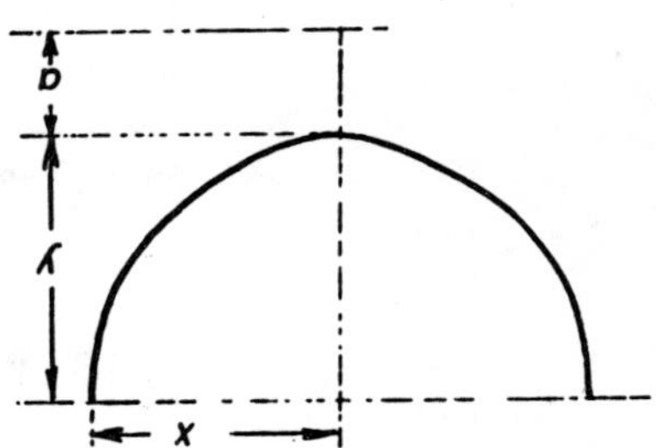

Solids

A6.13 Cube

volume $= a^3$
surface area $= 6a^2$

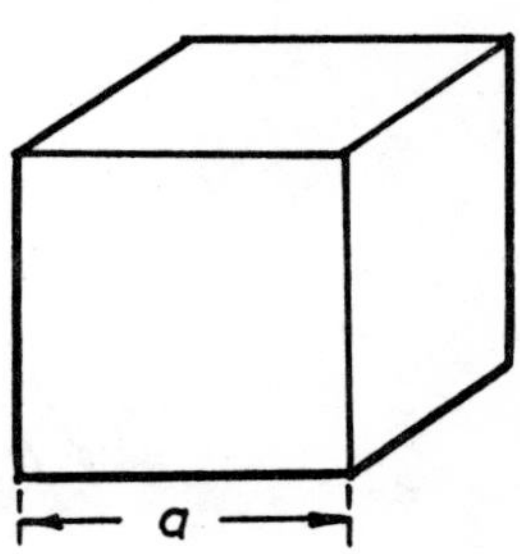

A6.14 Cylinder

volume = $0.7854D^2h$

surface area = πDh (without end surface)

= $\pi D(0.5D + h)$ (end surfaces included)

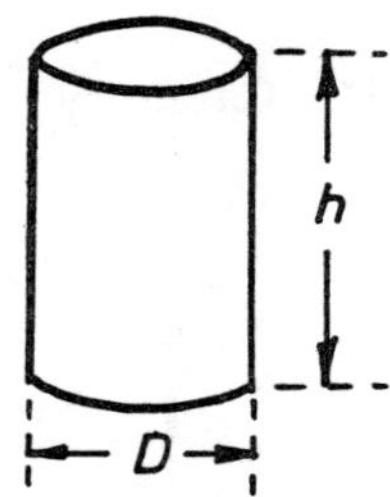

A6.15 Pyramid

volume = 1/3 (area of base)h

(Note: for area of base see **6.10**)

lateral area = 0.5(perimeter of base)s

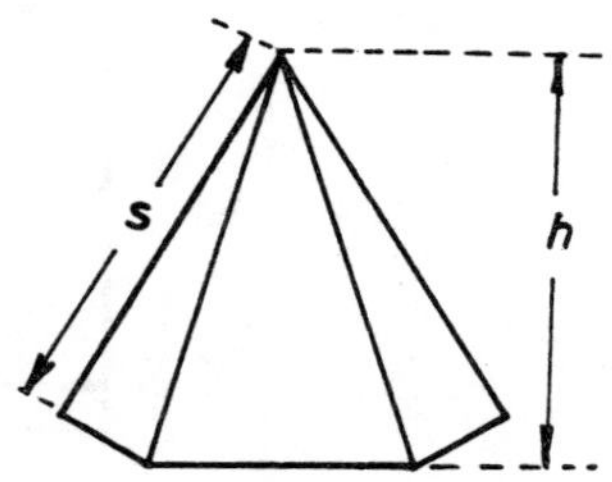

A6.16 Cone

volume $= 1.0472r^2h$

surface area $= \pi r(r^2 + h^2)^{1/2}$

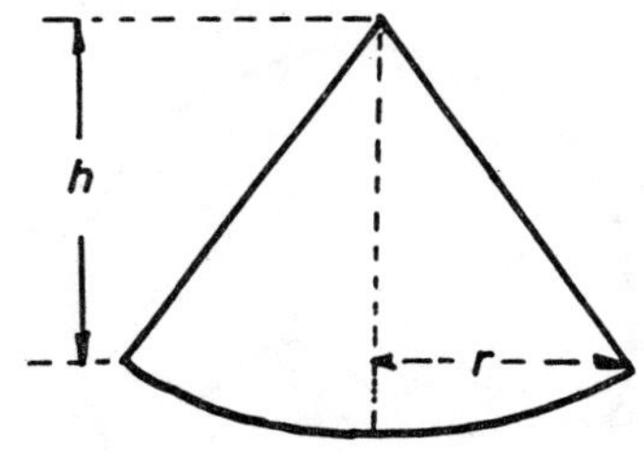

A6.17 Frustum of a Cone

volume

volume $= 1.0472h(r^2 + Rr + R^2)$

surface area $= \pi s(R + r)$

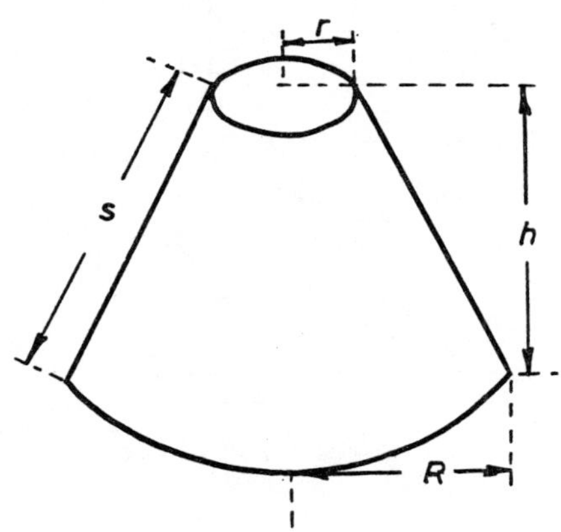

A6.18 Sphere

volume $= 0.5236D^3$
surface area $= \pi D^2$

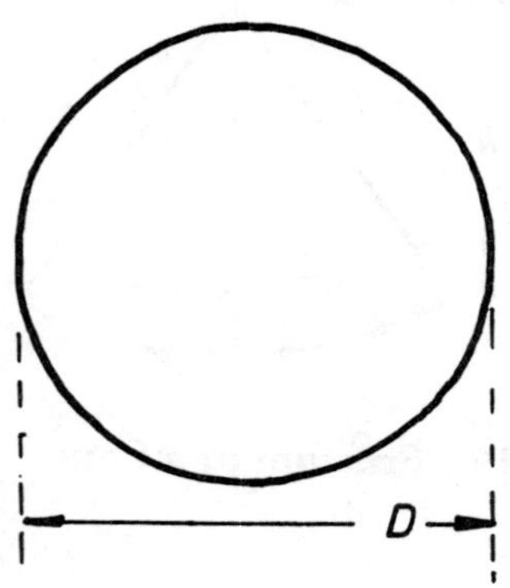

A6.19 Segment of a Sphere

volume $= \pi h^2 \left[\dfrac{c^2 + 4h^2}{8h} - \dfrac{h}{3}\right]$

spherical surface $= 0.7854(c^2 + 4h^2)$
total surface $= 0.7854(c^2 + 8rh)$

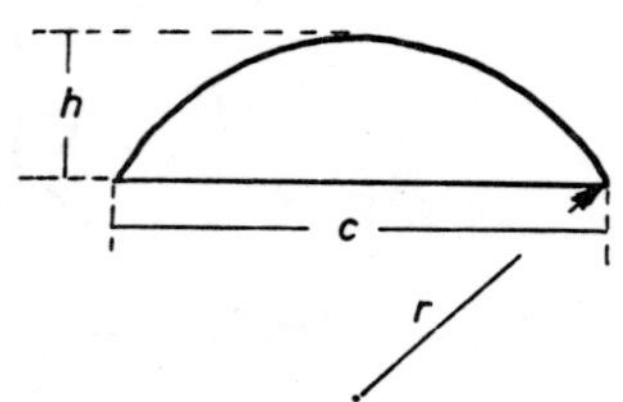

A6.20 Sector of a Sphere

volume = $2.0944r^2h$

total surface = $1.571r(4h + c)$

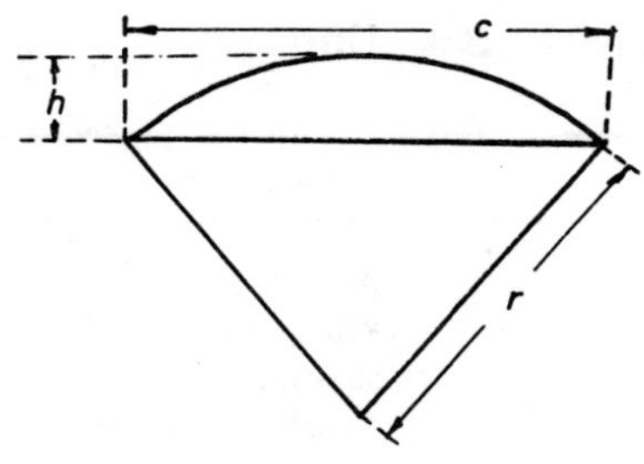

A6.21 Torus

volume = $0.25(\pi^2 d^2 D)$

surface = $D\pi^2 d$

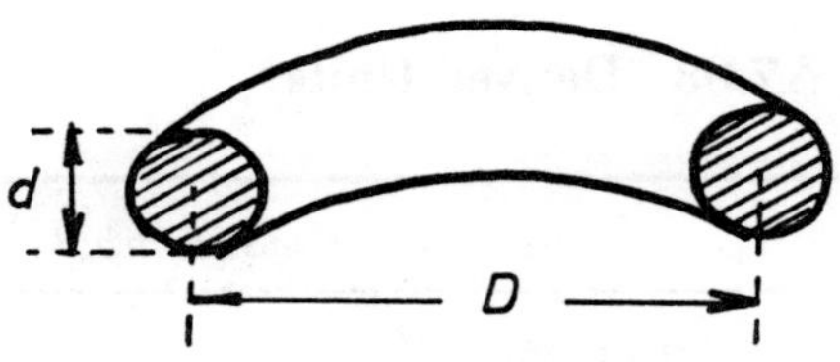

THE INTERNATIONAL SYSTEM OF UNITS (SI)

The following is a guide to familiarize the reader with the units, prefixes, symbols, and formulas used in the International System of Units. "SI" is the common language in which scientific and technical data will be presented worldwide in the not too distant future.

A7.01 Base Units

Quantity	*Unit*	*SI symbol*
length	meter	m
mass	kilogram	kg
time	second	s
electric current	ampere	A
thermodynamic temperature	Kelvin	K
amount of substance	mole	mol
luminous intensity	candela	cd

A7.02 Supplementary Units

plane angle	radian	rad
solid angle	steradian	sr

A7.03 Derived Units

Quantity	*Unit*	*SI symbol*	*Formula*
acceleration	meter per second^2		m/s^2
angular acceleration	radian per second^2		rad/s^2
angular velocity	radian per second		rad/s
area	square meter		m^2
density	kilogram per meter^3		kg/m^3
electric capacitance	farad	F	$A \cdot s/V$

(cont'd)

Quantity	*Unit*	*SI symbol*	*Formula*
electric field strength	volt per meter		V/m
electric conductance	siemens	S	A/V
electric inductance	henry	H	V · s/A
electric potential diff.	volt	V	W/A
electric resistance	ohm	Ω	V/A
electromotive force	volt	V	W/A
energy	joule	J	N · m
entropy	joule per kelvin		J/K
force	newton	N	$kg \cdot m/s^2$
frequency	hertz	Hz	(cycle)/s
illuminance	lux	lx	lm/m^2
luminance	candela per $meter^2$		cd/m^2
luminous flux	lumen	lm	cd·sr
magnetic field strength	ampere per meter		a/m
magnetic flux	weber	Wb	V · s
magnetic flux density	tesla	T	Wb/m^2
magnetomotive force	ampere	A	
power	watt	W	J/s
pressure	pascal	Pa	N/m^2
quantity of electricity	coulomb	C	A · s
quantity of heat	joule	J	N · m
radiant intensity	watt per steradian		W/sr
specific heat	joule per kg-kelvin		J/kg·K
stress	pascal	Pa	N/m^2
thermal conductivity	watt per m-kelvin		W/m·K
velocity	meter per second		m/s
viscosity, dynamic	pascal-second		Pa·s
viscosity, kinematic	m^2 per second		m^2/s
voltage	volt	V	W/A
volume*	cubic meter		m^3
wavenumber	reciprocal meter		(wave)/m
work	joule	J	N · m

* In normal engineering work, where high precision is not required, the use of the liter as a unit to express volume is acceptable.

A7.04 Prefixes

Multiplication factor	*Prefix*	*SI Symbol*
1 000 000 000 000 = 10^{12}	tera	T
1 000 000 000 = 10^{9}	giga	G
1 000 000 = 10^{6}	mega	M
1 000 = 10^{3}	kilo	k
100 = 10^{2}	hecto	h
10 = 10^{1}	deca	da
0.1 = 10^{-1}	deci	d
0.01 = 10^{-2}	centi	c
0.001 = 10^{-3}	milli	m
0.000 001 = 10^{-6}	micro	μ
0.000 000 001 = 10^{-9}	nano	n
0.000 000 000 001 = 10^{-12}	pico	p

Examples

2460 m or 2.46×10^{3} m is written as 2.46 km
34,000 W or 34.0×10^{3} W is written as 34.0 kW
0.0046 V or 4.6×10^{-3} V is written as 4.6 mV
6,300,000 W or 6.3×10^{6} W is written as 6.3 MW

Section B

CONVERSION FACTORS

B1 Linear Measures

Multiply	*by*	*to obtain*
mm	0.1	cm
mm	0.03937	in.
mm	0.003281	ft
cm	0.01	m
cm	0.3937	in.
cm	0.03281	ft
m	1.093	yds
m	3.281	ft
km	0.6213	miles, statute
km	1000	m
miles, statute	1760	yds
miles, statute	0.8684	miles, nautical
miles, statute	1.609	km
miles, nautical	1.1515	miles, statute
miles, nautical	1853	m
in.	25.4	mm
in.	2.54	cm
ft	30.479	cm
ft	0.3048	m
yd	0.9149	m
yd	3	ft

B2 Weights

Multiply	*by*	*to obtain*
g	15.43	grains
g	0.03502	oz
kg	2.2046	lb
kg	35.274	oz
metric ton	1.1023	short ton
metric ton	2240	lb
grains	0.0648	g
oz	0.0625	lb
oz	437.5	grains
oz	28.35	g
lb	0.4536	kg
lb	16	oz
lb	7000	grains
lb	453.59	g
short tons	0.9072	metric tons
g	980.66	dynes
lb	444820.0	dynes

B3 Area

Multiply	*by*	*to obtain*
cm^2	0.00108	ft^2
cm^2	0.155	$in.^2$
cm^2	100	mm^2
m^2	10.76	ft^2
m^2	1.196	yd^2
km^2	0.3861	$miles^2$
$in.^2$	6.45	cm^2
ft^2	0.0929	m^2
ft^2	144	$in.^2$
ft^2	929.03	cm^2
yd^2	9	ft^2
yd^2	0.8361	m^2
$mile^2$	640	acre, US
$mile^2$	2.59	km^2
acre, US	43560	ft^2

B4 Volume

Multiply	*by*	*to obtain*
cc	0.001	l
cc	1	ml
cc	0.03381	oz, fluid, US
cc	0.06102	$in.^3$
l	0.26418	gal, US
l	0.03532	ft^3
m^3	1.3079	yd^3
m^3	35.314	ft^3
m^3	264.17	gal, US
m^3	1000	l
gal, US	128	fl. oz.
gal, US	0.13368	ft^3
gal, US	3.7854	l
$in.^3$	0.5541	fl. oz.
$in.^3$	0.01639	l
$in.^3$	16.387	ml
ft^3	7.481	gal, US
ft^3	0.02832	m^3
ft^3	28.316	l
gal, US	0.003785	m^3
l	0.001	m^3
yd^3	0.7646	m^3

B5 Specific Weights and Volumes

Multiply	*by*	*to obtain*
kg/m^3	0.9990	oz/ft^3
kg/m^3	0.06243	lb/ft^3
m^3/kg	1.001	ft^3/oz
m^3/kg	16.02	ft^3/lb
g/m^3	0.435	$grains/ft^3$
g/cc	8.3452	lb/gal, US
g/cc	62.428	lb/ft^3
ppm	0.0584	grains/gal, US
$grains/ft^3$	2.299	g/m^3
oz/ft^3	1.001	kg/m^3
lb/ft^3	16.02	kg/m^3
lb/ft^3	0.01602	g/cc
lb/gal, US	0.11983	g/cc
lb/gal, US	7.48	lb/ft^3
lb/gal, US	0.11983	kg/l
ft^3/lb	0.06243	m^3/kg
ft^3/lb	62.4262	l/kg

B6 Flow Rates and Speeds

Multiply	*by*	*to obtain*
m^3/s	2118.9	ft^3/min
m^3/min	0.5886	ft^3/s
m^3/h	4.4028	gal, US/min
ft^3/min	0.47193	lb/s
ft^3/min	0.02832	m^3/min
ft^3/s	448.83	gal, US/min
cm/s	1.9685	ft/min
ft/min	0.508	cm/s
ft/min	0.018288	km/h
ft/min	0.3048	m/min
ft/min	0.011364	miles/h
gal/min	0.002228	ft^3/s
gal/min	0.22712	m^3/h
gal/min	0.063088	l/s
l/min	0.0005886	ft^3/s
l/min	0.0044028	gal/s
km/h	54.68	ft/min
km/h	0.9113	ft/s
km/h	0.6214	miles/h
knots	1.853	km/h
m/s	3.2808	ft/s
miles/h	88	ft/min
miles/h	1.609	km/h
miles/h	0.8684	knots
ft^3/min	0.0004719	m^3/s
gal/min	0.00006309	m^3/s

B7 Pressure

Multiply	*by*	*to obtain*
atm.	760	mm of Hg
atm.	14.696	psi
atm.	29.921	in. Hg
atm.	33.899	ft of H_2O
atm.	76	cm of Hg
cm of Hg	0.013158	atm.
cm of Hg	0.44604	ft of H_2O
cm of Hg	0.19337	psi
ft of H_2O	0.029499	atm.
ft of H_2O	2.242	cm of Hg
ft of H_2O	0.43352	psi
in. of H_2O	0.0024583	atm.
in. of H_2O	0.1868	cm of Hg
in. of Hg	345	mm of H_2O
lb/ft^2	0.035913	cm of Hg
lb/ft^2	4.88	kg/cm^2
kg/cm^2	14.223	psi
psi	5.1715	cm of Hg
psi	68,947	$dynes/cm^2$
psi	2.3066	ft of H_2O
psi	2.0360	in. of Hg
psi	27.673	in. of H_2O
psi	0.0703	kg/cm^2

(see appendix for conversion factors to change to SI (pascal).

B8 Work, Power, and Force

Multiply	*by*	*to obtain*
ft-lb	0.1383	m-kg
ft-lb	0.001285	Btu
ft-lb	0.000324	kcal
ft-lb	0.00003030	mechanical hp
ft-lb/s	0.001818	mechanical hp
in.-lb	0.01152	m-kg
ft-lb/ft^3	4,883	m-kg/m^3
hp h	273,700	m-kg
mechanical hp	33,000	ft-lb
m-kg	7.231	ft-lb
m-kg	86.81	in.-lb
m-kg	3.654×10^{-5}	hp h
m-kg/m^3	0.2048	ft-lb/ft^3
Btu/s	1,054	W
Btu/min	17.57	W
cal/s	4.184	W
cal/min	0.06973	W
erg/s	1.0×10^{-7}	W
ft-lb force/h	3.766×10^{-4}	W
hp (electrical)	746.0	W
dyne	0.00001	N
kg-force	9.807	N
lb-force (Av)	4.448	N

B9 Power

Multiply	*by*	*to obtain*
Btu/s	1.055	kW
Btu/s	907.3	kcal/h
Btu/h	0.0003931	hp
Btu/h	0.000293	kW
hp	0.7459	kW
hp	745	W
hp	641.2	kcal/h
hp	2544	Btu/h
hp	550	ft-lb/s
kW	1000	W
kW	1.34	hp
kW	860	kcal/h
kW	3413	Btu/h
W	0.001	kW
W	0.001342	hp
W	0.860	kcal/h
W	3.413	Btu/h

B10 Heat

Multiply	*by*	*to obtain*
Btu	0.0029	kWh
Btu	0.252	kcal
Btu	107.6	m-kg
Btu/lb	0.5556	kcal/kg
Btu/short ton	0.000278	kcal/kg
Btu/in.^{2}	390.8	kcal/m^2
Btu/ft^2	2.713	kcal/m^2
Btu/ft^3	8.90	kcal/m^3
kWh	860	kcal
kWh	3,441	Btu
kcal	3.97	Btu
kcal/kg	1.7999	Btu/lb
kcal/kg	3,599.8	Btu/short ton
kcal/m^2	0.002559	Btu/in.^{2}
kcal/m^2	0.3686	Btu/ft^2
kcal/m^3	0.1124	Btu/ft^3
Btu	1,055.056b	J (joule)
Btu/in./s/ft^2 °F	518.9	W/m·°K
Btu/ft^2	11,350	J/m^2
cal/cm^2	41,840	J/m^2
kcal/kg	4,184	J/kg
kcal/kg°C	4,184	J/kg·°F

B11 Pressure Conversion to SI Unit

Multiply	*by*	*to obtain*
atm.	1.013×10^5	Pa
bar	1.000×10^5	Pa
dyne/cm^2	0.100	Pa
g(force)/cm^2	98.07	Pa
in. of Hg (60°F)	3377.0	Pa
in. of H_2O (60°F)	248.8	Pa
mm of Hg (0°C)	133.3	Pa
psf	47.88	Pa
psi	6895.0	Pa

B12 Viscosity

Multiply	*by*	*to obtain*
ft^2/s	0.0929	m^2/s
P	0.1	Pa/s
lb(force)s/ft^2	47.88	Pa/s
stokes	0.0001	m^2/s

B13 Metric Standard Units

Multiply	*by*	*to obtain*
Kp (Kilopound)	70.93164	pdl (poundal)
Kp	2.20462	lb (pound)
Kp/cm^2 (at)	14.22334	p.s.i.
bar	14.50377	p.s.i.
mm QS (Torr)	0.01933676	p.s.i.
N/m	1.4503767×10^{-4}	p.s.i.
Kp/mm^2	1422.334	p.s.i.
Kp/m^2	0.0014223	p.s.i.

Multiply	*by*	*to obtain*
Kg/m	0.671969	lbs./ft
Kg/m^2	0.204816	$lbs./ft^2$
Kg/cm^2	36.1273	$lbs./in.^3$
Kg/l	62.4266	$lbs./ft^3$

Multiply	*by*	*to obtain*
$Nm^3{}_{tr}$	37.228	SCF_{Dry}
$Nm^3{}_{tr}$	37.889	SCF_{Moist}
$Nm^3{}_f$	37.660	SCF_{Moist}
$Nm^3{}_f$	37.004	SCF_{Dry}
$Nm^3{}_{tr}/Kg$	16.886	$SCF_{Dry}/lb.$
$Nm^3{}_f$	17.082	$SCF_{Moist}/lb.$
$Kg/Nm^3{}_{tr}$	0.05922	$Lb./SCF_{Dry}$
$Kg/Nm^3{}_f$	0.058537	$Lb./SCF_{Moist}$
$Kcal/Nm^3{}_{tr}$	0.1066	Btu/SCF_{Dry}
$Kcal/Nm^3{}_f$	0.10537	Btu/SCF_{Moist}
$KJ/Nm^3{}_{tr}$	0.44631	Btu/SCF_{Dry}
$KJ/Nm^3{}_f$	0.44116	Btu/SCF_{Moist}

Standard conditions for Metric System (N) = 0 C, 760 mm Hg.
Standard conditions for English System = 60 F, 30 in. Hg.

REFERENCES

(1) Kuehl, H. 1929. *Zement* 18:833.

(2) Lea, F. M. and Desch, C. H. 1935. *The chemistry of cement and concrete.* Longmans, Green & Co.

(3) Peray, K. E. and Waddell, J. J. 1972. *The rotary cement kiln.* New York: Chemical Publishing Co.

(4) Boque, R. H. 1929. Industrial Engineering Chemistry. *Anal. Ed.* 1:192.

(5) Duda, W. H. *Cement Data Book.* Bauverlag GmbH Wiesbaden.

(6) Denver Equipment Co. *Dry solids in solution.* Bulletin S1C-B4.

(7) Okorokov, S. D. 1952. *Technologia Viashuschich Vieshtchestv.* Moscow.

(8) Duda, W. H. *Cement Data Book.* Bauverlag GmbH Wiesbaden.

(9) U. S. Bureau Mines Tech. Paper. 1927. 384.

(10) Allis-Chalmers, Milwaukee, Wisconsin, Bulletin No. 22 B 1212.

(11) Schwiete, H. E. 1932. *The specific heats of Portland cement clinker.* Tonindustrie Zeitung.

(12) Fed. Register. Dec. 1971. *Standard Performance for Stationary Sources.*

(13) F. C. Bond. May 1959. Third theory of comminution. *Mining Engineering.*

(14) F. C. Bond. May 1958. Grinding ball size selection. *Mining Engineering.*

(15) Giesking, D. H. Jaw crusher capacities. *Mining Transactions.* Vol. 184.

(16) *Perry's Chemical Engineers Handbook.* 4th Ed. 9–43. New York: McGraw-Hill.

(17) Keenan, J. H. and Keyes, F. G. 1936. *Thermodynamic properties of steam.* New York: John Wiley & Sons.

(18) Schwiete, H. E. and von Gronow, E. Specific heat of cement raw mixes. *Zement* 24.

INDEX

D

E

F

G

I

J

K

L

M

N

O

P

Q

R

S

T